The BeagleBone Black Primer

Brian McLaughlin

Que®

The BeagleBone Black Primer

ISBN-13: 978-0-7897-5386-1
ISBN-10: 0-7897-5386-3

Library of Congress Control Number: 2015946119

Printed in the United States of America

First Printing: September 2015

Trademarks

All terms mentioned in this book that are known to be trademarks or service marks have been appropriately capitalized. Que Publishing cannot attest to the accuracy of this information. Use of a term in this book should not be regarded as affecting the validity of any trademark or service mark.

This book was not created by and is not endorsed by Texas Instruments.

Warning and Disclaimer

Every effort has been made to make this book as complete and as accurate as possible, but no warranty or fitness is implied. The information provided is on an "as is" basis. The author and the publisher shall have neither liability nor responsibility to any person or entity with respect to any loss or damages arising from the information contained in this book.

Special Sales

For information about buying this title in bulk quantities, or for special sales opportunities (which may include electronic versions; custom cover designs; and content particular to your business, training goals, marketing focus, or branding interests), please contact our corporate sales department at corpsales@pearsoned.com or (800) 382-3419.

For government sales inquiries, please contact governmentsales@pearsoned.com.

For questions about sales outside the U.S., please contact international@pearsoned.com.

Editor-in-Chief
Greg Wiegand

Executive Editor
Rick Kughen

Development Editor
Ginny Bess Munroe

Managing Editor
Sandra Schroeder

Project Editor
Seth Kerney

Copy Editor
Bart Reed

Indexer
Tim Wright

Proofreader
Laura Hernandez

Technical Editor
Anton Olsen

Publishing Coordinator
Kristen Watterson

Interior Designer
Mark Shirar

Cover Designer
Mark Shirar

Compositor
Jake McFarland

Photography
Helene McLaughlin

Contents at a Glance

Introduction 1

CHAPTER 1 Embedded Computers and Electronics 5

CHAPTER 2 Introduction to the Hardware 15

CHAPTER 3 Getting Started 25

CHAPTER 4 Hardware Basics 39

CHAPTER 5 A Little Deeper into Development 57

CHAPTER 6 Trying Other Operating Systems 71

CHAPTER 7 Expanding the Hardware Horizon 81

CHAPTER 8 Low-Level Hardware and Capes 97

CHAPTER 9 Interacting with Your World, Part 1: Sensors 113

CHAPTER 10 Remote Monitoring and Data Collection 127

CHAPTER 11 Interacting with Your World, Part 2: Feedback and Actuators 149

CHAPTER 12 Computer Vision 171

CHAPTER 13 Sniffing Out Car Trouble 189

CHAPTER 14 Ground Control to Major Beagle 205

CHAPTER 15 Moving Forward 225

Index 233

Table of Contents

Introduction **1**
Who This Book Is For 1
How This Book Is Organized 2
Conventions Used in This Book 3
Let Me Know What You Think 3

Chapter 1 Embedded Computers and Electronics **5**
What Are Embedded Electronics? 5
Arduino 9
What Should Readers Get Out of This Book? 12

Chapter 2 Introduction to the Hardware **15**
A Short Lineage of the BeagleBone Black 15
BeagleBone Black Hardware Specification 19
Processor 20
RAM 21
Onboard Flash and MicroSD External Storage 22
Ethernet 22
General-Purpose Input/Output 22

Chapter 3 Getting Started **25**
Setting Up and Saying "Hello, World!" 26
Connecting to Ethernet 32

Chapter 4 Hardware Basics **39**
Electronics Basics: Voltage, Current, Power, and Resistance 39
The Short Circuit 43
The Resistor 45
Diodes and LEDs 48
Build an LED Circuit 50

Chapter 5 A Little Deeper into Development **57**
Interpreted Code 57
Python—A Step Above Interpreted Language 59
Implementing Blinking Lights In Python 62
Compiled Code 65

Chapter 6 Trying Other Operating Systems 71
History of the Linux World: Part I 71
Picking an Operating System 73
Loading the microSD Card 73
Chapter 7 Expanding the Hardware Horizon 81
Binary Basics 81
Hardware Representation 83
Serial Communications 91
Inspecting UART 93
Chapter 8 Low-Level Hardware and Capes................................. 97
Linux Hardware Through The File System...................................97
Hardware in the File System..100
One Pin, Multiple Functions 103
Hardware Configuration 108
Chapter 9 Interacting with Your World, Part 1: Sensors 113
Sensor Basics 113
Analog Versus Digital 120
Sample Rates 124
Chapter 10 Remote Monitoring and Data Collection.................. 127
Project Outline 127
Wiring Up The Project 130
Seeing the Light..134
Publishing the Sensor Data...137
Start Collecting Data 142
Chapter 11 Interacting with Your World, Part 2: Feedback and Actuators 149
Controlling Current 149
Blinking to Fading 156
Vibration Motors...159
Servo Motors 161
Stepper Motors 165
Chapter 12 Computer Vision 171
Connecting a Camera 171
Utilizing OpenCV Libraries..177

A Better Photo Booth 178
Cascade Classifiers 180
Tracking a Face 182
Chapter 13 Sniffing Out Car Trouble 189
Car Computers 189
Interfacing to the Car 191
Reading the Car's Status 198
Interpreting the Data 199
Chapter 14 Ground Control to Major Beagle 205
Radio Data 205
WiFi 210
Software Defined Radio 212
Grabbing Libraries with Git 215
Radio Testing 216
Calibrating the Radio 219
Listening to Aviation Data 221
BeagleBone Black Air Traffic Control Station 223
Chapter 15 Moving Forward 225
Project Ideas 226
Portable Gaming Solutions 226
Weather Station 227
In-Car Computer 227
More Advanced Aircraft "RADAR" 228
Satellite Ground Station 228
Tools 230
Resources 230
Index 233

About the Author

Brian McLaughlin is an engineer by profession and by hobby. Brian earned a bachelor's degree in computer science from North Carolina State University and a masters of engineering in systems engineering from the University of Maryland. With a solid foundation in software, Brian was initially exposed to more advanced topics in hardware while working on the Hubble Space Telescope Project. Over time, Brian began writing for GeekDad and has become a part of the growing Maker community. Brian lives in Maryland with his beautiful wife and two boys.

Dedication

For Mom & Dad

Acknowledgments

I wish I could acknowledge everyone who ever taught me something about STEAM (science, technology, engineering, art, and mathematics) topics, but that would be almost every teacher, instructor, mentor, and co-worker I have ever had to this point in my life. I would like to start by thanking the Integration and Test and software development teams for the Wide Field Camera 3 instrument on the Hubble Space Telescope, the first place I worked where the rubber met the road between hardware and software. I would like to thank my mentors—specifically Larry Barrett and Curtis Fatig—with whom I worked on the James Webb Space Telescope project and other projects. From them I was always learning something about engineering, working in a high-pressure environment, travelling the world, and finding out about life in general. I would like to thank my friends at GeekDad who helped me find a passion for writing about technical topics for fun and not just for my 9-to-5 job.

I would like to thank the people and companies who provided hardware and parts in support of this book including Tektronix, Oscium, SparkFun, and Element14.

I would like to offer my apologies to my neighbors in my little cul-de-sac in Columbia, Maryland. Writing a book while still holding down a full-time job was much harder than I had anticipated, and my lawn and yard suffered as a consequence. I promise I will keep them looking better!

I of course need to thank my parents, Glen and Diane, and my brother, Glen. Our parents always encouraged us to explore, learn, and grow, and my brother, in addition to sharing systems with me, showed me the Mosaic web browser before most people knew what the Web even was. I also need to thank my Uncle Lou, who passed along computers as he upgraded, and always made sure we were working on learning the basics of flying with *Flight Simulator*. It was also thanks to my parents and my Uncle Lou that I went to Space Camp in seventh grade.

Finally, I must acknowledge all of the rest of my loving family, particularly my beautiful wife, Helene, and my boys, Sean and Liam. Without everyone's support, patience, understanding, love, and patience, I never would have finished this book. (Yes, patience needed to be there twice.)

We Want to Hear from You!

As the reader of this book, *you* are our most important critic and commentator. We value your opinion and want to know what we're doing right, what we could do better, what areas you'd like to see us publish in, and any other words of wisdom you're willing to pass our way.

We welcome your comments. You can email or write to let us know what you did or didn't like about this book—as well as what we can do to make our books better.

Please note that we cannot help you with technical problems related to the topic of this book.

When you write, please be sure to include this book's title and author as well as your name and email address. We will carefully review your comments and share them with the author and editors who worked on the book.

Email: feedback@quepublishing.com

Mail: Que Publishing
ATTN: Reader Feedback
800 East 96th Street
Indianapolis, IN 46240 USA

Reader Services

Visit our website and register this book at quepublishing.com/register for convenient access to any updates, downloads, or errata that might be available for this book.

Introduction

The world is becoming a place where the traditional technical disciplines of science, technology, engineering, and mathematics (STEM) have boiled over into the world of art to produce STEAM (that is, STEM with Art thrown in the mix). It is the beginning of a new Renaissance. Just like in Da Vinci's time, cross-domain studies in all the STEAM topics are critical, and they are often unified via some form of electronics.

For example, an art installation may allow for a mechanical, interactive sculpture. This sculpture may require some "senses" to understand changes in the environment—everything from a change in temperature to the traditional senses of touch, sight, sound, taste, and smell. These changes are processed by some electronic means, and then some action is taken with that information. Maybe the head of a statue "looks" at you when you walk past.

For some, even the most technical work can seem like art. The layout of a circuit on a board, an elegant programming solution, and the RS-25 rocket engine, among many other technical solutions, are all like looking at art to me.

This book strives to provide the information necessary for you to find your own art in the world of STEAM. We will use a very accessible and powerful electronics board for this task—the BeagleBone Black.

Who This Book Is For

Targeting an audience for a book such as this can be tricky. For example, there are people in the artistic world—I've known many myself—who want to integrate electronics into their art projects, but they find the task daunting. I wrote this book so that those individuals can start to understand how electronics work and forge a path forward to bring their artistic creations to life.

There are others out there who have plenty of experience with electronics and building projects, but want to move toward using the power of the BeagleBone Black. For those readers, many of the sections of this book will provide quick reference to information on accessing the pins and functionality of the board and how to accomplish the basic tasks they need to know to build larger projects.

None of the projects are completely finished out in the course of the book. They are left at the level of the breadboard and still lack the finalized look of a project with a completed enclosure and installation. This is on purpose. I only give you enough information to be dangerous—and to go out on your own and build something amazing.

How This Book Is Organized

I try not to make any assumptions concerning what you may know about electronics and computers, other than basic familiarity with traditional desktop environments. With that in mind, I attempt to start off slowly. Here's what you'll find in the chapters of this book:

- **Chapters 1–5:** These chapters provide you with an overview of embedded electronics and development platforms. In these chapters, you learn what the BeagleBone Black represents and what major parts it is made from. You also learn how to procure a board, hook it up for the first time, and get something running. You learn some basics of electronics and how to get electrons to obey your will and desires. Well, they won't obey, but they will follow the paths you force them down and at the rates you desire. Finally, you learn how to use programming to make things happen on the board, and you get some exposure to a couple of the programming languages available and learn some of the differences between them.
- **Chapters 6–8:** These chapters provide some more advanced topics on hardware interactions and the environments you can use for operating your board. A number of operating system environments are available to run on the board, and in these chapters you learn how to switch out an operating system. You also learn some more low-level hardware interface information and about the ecosystem of standardized hardware expansion boards for the BeagleBone Black (known as Capes).
- **Chapters 9–14:** These chapters offer some insight into building more complex projects with the BeagleBone Black. You learn about how sensors work, build an environment-monitoring station for your working area, and find out how to manipulate items in your environment via motors. Finally, you get into various projects. You'll learn how to give your creator's vision, which can be used to actually track a person's face. You'll also tap into the computer in your car, and even listen to the data sent from aircraft so that you can track the aircraft in your area.
- **Chapter 15:** This chapter leaves you with some room to think about where you can go after reading the book. You'll learn about places to start expanding on what you have learned, a little about securing your BeagleBone Black, and some of my favorite project resources.

Only you know how much knowledge and experience you have as you begin reading this book. If you're completely new to this world, you will probably want to work through the book sequentially and build your knowledge as you go. If you're an experienced user who just wants examples of how to accomplish specific tasks with the BeagleBone Black, you will probably tend to bounce around and grab the bits you need.

Conventions Used in This Book

A couple of conventions are used throughout this book. Monospaced font calls out source code and terminal interactions. For example:

```
print "This is a line of source code"
```

```
~/bbb-primer/$ this is a terminal interaction
```

Note that source code is called out as a Listing, but terminal interactions are not. Also note that, in terminal interaction, content the user should enter is in **`bold mono font`**.

Let Me Know What You Think

If you want to contact me, feel free to email me at **bjmclaughlin@gmail.com**. I welcome questions that clarify points made in the book and constructive criticism.

However, as with many technical topics, there is often more than one way to accomplish a goal. Therefore, if you offer a comment that just shows a different way to accomplish the same thing or a quicker, more efficient way when the point of the example was obviously to be clear and thoughtful and not efficient, I will likely mentally acknowledge your point, archive the email, and move on with my life. Remember, my goal is to be as clear and accessible to as many levels of interest as possible.

With all of that said, I think you will enjoy the book, and I hope that you will learn how to accomplish whatever it is you set out to do after reading it. *Allons-y!*

Embedded computers and embedded electronics have been around for a long time but with the increasing availability of small computing platforms, the ability to build complex projects is a skill within the reach of more and more hobbiests. This book will introduce you to embedded computing through one of these platforms, the BeagleBone Black.

What Are Embedded Electronics?

You might have followed one of many paths that brought you to pick up this book. You might have seen a demonstration of the BeagleBone Black in a project write-up in an online forum, or perhaps you saw the BeagleBone Black in action at a Maker Faire or Mini-Maker Faire. Generally, the BeagleBone Black is not intended to replace your desktop or laptop computer. Instead, it is intended to be used as an embedded computer for a project. An *embedded computer* is a computer that is specifically part of your project and goes with the project when it moves away from your workbench.

For example, let's say you develop a project using a desktop or laptop computer. When you complete the project, you might unplug from the computer you used for development and plug your project into another computer. An embedded computer, however, remains with the project. Your project may still require a connection to an outside computer or utilize other computing resources, but that part of the project that is always a part of the build is embedded.

A laptop or a desktop may be embedded in a project as a stand-alone platform. Many robotic projects utilize a powerful laptop or desktop-grade computer to operate, as shown in Figure 1.1. Generally, however, you want something more lightweight that can run on direct current (DC) power rather than wall power, and you want something that provides a number of general-purpose input/output (GPIO) ports. This is where a platform such as the BeagleBone Black comes in.

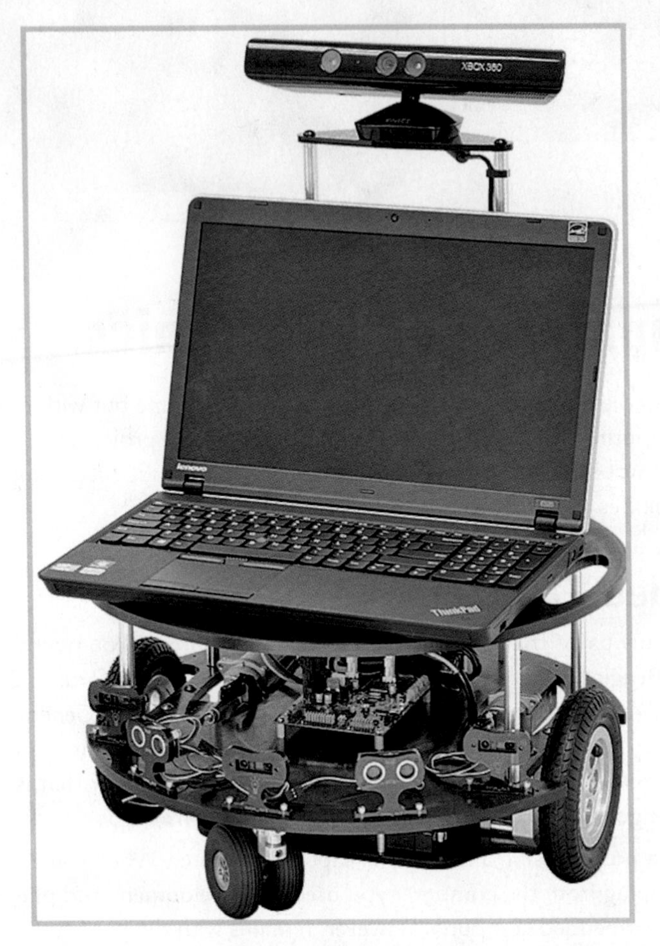

FIGURE 1.1 A retired robotics platform from Parallax that can utilize a laptop as an embedded part of the robot.

The BeagleBone Black is a small-footprint computing platform that packs a computational kick. How small of a footprint? Figure 1.2 provides a comparison of a laptop, an ATX motherboard footprint, a Micro ATX footprint, and the BeagleBone Black.

It is important to remember that you do make some trade-offs in power and performance by going to a smaller embedded footprint. All of the other sizes can support cutting-edge (at least at the time of this writing) Intel i7 processors that run well over a 3GHz clock and have a ton of RAM. If you're not sure what that means, here are a couple of rules of thumb:

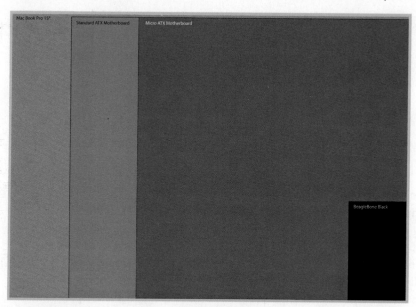

FIGURE 1.2 A comparison of relative sizes of a laptop, an ATX motherboard, a micro ATX motherboard, and the BeagleBone Black.

- The *clock* on a computer is used to orchestrate how fast everything happens inside the processor. The faster the clock, measured in cycles per second, the more instructions the computer can execute in that second.
- *RAM,* which stands for *random access memory,* is where your programs store the variables and other information they use during execution, so the more RAM you have onboard, the more "room" your software has to run in. If your executing program runs out of room, at best, some data will have to move to storage that is much slower than RAM. At worst, your program will not work at all.

Table 1.1 shows a comparison of some basic specifications between a laptop, a desktop, and a BeagleBone Black.

TABLE 1.1 Comparison of Basic Laptop, Desktop, and BeagleBone Black Specifications

	Volume (in³)	Number of Processor Cores	Clock Speed (GHz)	RAM (GB)	Weight (lbs.)	Power Consumption (W)
MacBook Pro 15"	98	4	2.0	8	4.46	85
ASUS Core i7	1590	4	3.1	8	17.4	100
BeagleBone Black	1.3	1	1	0.5	0.09	5

As you can plainly see, the BeagleBone Black isn't a replacement for any current-generation major computing platform. However, picture yourself building a project, such as the aircraft-tracking project from this book. The aircraft-tracking project is intended to be a portable rig with an antenna for tracking aircraft in the local area. If you wanted to deploy it as a standalone system, you would want something that's small and lightweight, with low power requirements. It needs to be small enough for easy weatherproofing, and yet have enough power to decode the data packets from an aircraft and either display them or transmit them over a network. This is where a purpose-built embedded device such as the BeagleBone Black shines.

Another powerful utility in embedded systems is the capability to directly control other electronics. If you wanted to build your own electronics for your project (which we will do in this book), and you used a laptop or desktop machine, you would need to make sure your electronics could talk to a USB port on your machine. In the past, there were more options with serial ports and parallel ports, but most modern systems use only a USB port and will need additional hardware to access serial or parallel data.

The BeagleBone Black provides a whole slew of direct GPIO ports that allow you to directly control or read the voltage on a wire. That is huge! Let's say, for simplicity's sake, that all you wanted to do is make a light-emitting diode (LED) light up. From the BeagleBone Black, all you need to do is connect the LED through a resistor into a GPIO socket and then into a ground connection on the board. It's that simple! This hookup is shown in Figure 1.3. If you don't understand what you see here, don't worry—it's explained later.

From those simple roots, you have the option to manipulate and read data on all kinds of protocols or define your own. The best part is, if you do have a USB device you want to use in your project, the BeagleBone Black has USB ports and many other ports just like you would find on any laptop or desktop.

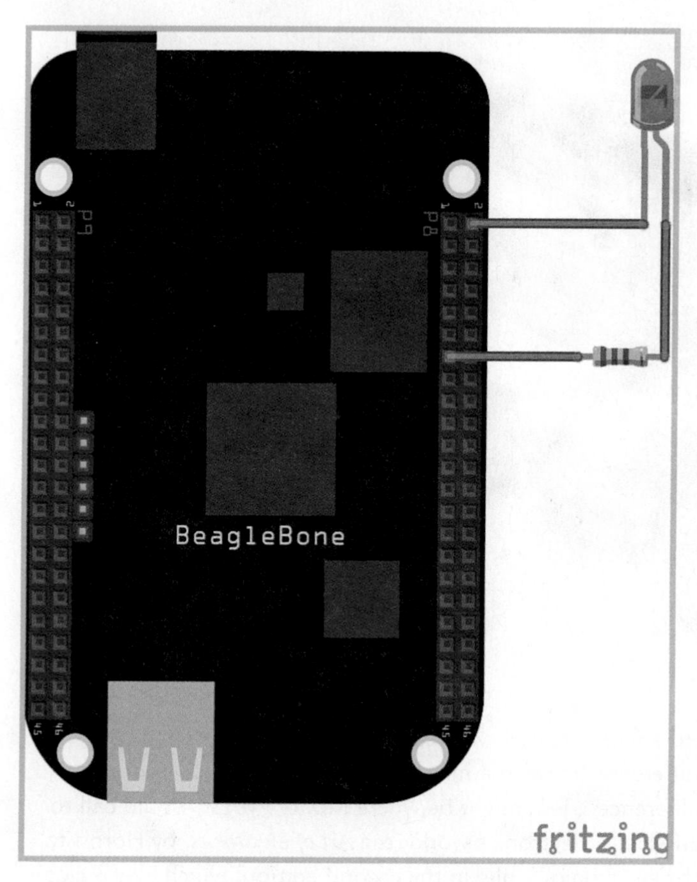

FIGURE 1.3 Easy setup for lighting up an LED.

Arduino

Another popular platform out there right now for embedded systems is the Arduino (see Figure 1.4). A few crucial differences between the Arduino and the BeagleBone Black are worth highlighting. They aren't competing platforms; they are complementary and can be used together to level up your project's capabilities.

FIGURE 1.4 The Arduino Uno, Revision 3.

The Arduino runs what is called a *microcontroller,* whereas the BeagleBone Black runs a *microprocessor.* So what is the difference between a microcontroller and a microprocessor? To the layman, there is little difference. (This might be where Monty Python would call to "consult the Book of Armaments!") In electronics world, the *Art of Electronics,* by Horowitz and Hill, is often considered the electronics bible. In the second edition, which I keep close by my side at my desk, Chapter 8 discusses chips that are considered both microcontrollers and microprocessors, citing the Intel 80386 as a "powerful number cruncher." The general idea is that a microprocessor is usually much more powerful and optimized for pure processing power, and it is up to the user to include everything else. On the other hand, according to Horowitz and Hill, microcontrollers are:

> Single chip processors with various input, output, and memory functions included on the same chip, for stand alone use...intended as a dedicated controller in an instrument, rather than as a versatile computation device.

The microcontroller at the heart of the Arduino is the Atmel ATMega 328P. This is a great microcontroller that makes rapid development easy. Generally, in the microcontroller environment, if you boot the board without the software you've already loaded, the board will do nothing. When you write code for a microcontroller setup like the Arduino, you write "bare-metal to the controller" with nothing helping you and nothing hindering you. It's just you and the hardware environment—and that is a freeing sensation! This means you have absolute control over what happens on the board, and the hardware world is your oyster!

On the BeagleBone Black, with its microprocessor, your programs run within an operating system. On your desktop or laptop, you can run Windows, OS X, or Linux. The BeagleBone Black supports a number of Linux distributions, and people are working to port other open-source operating systems all the time. So, why would you want to have an operating system getting between you and your pristine hardware? Aside from the raw increase in power, there are two good reasons:

- **To simplify a powerful set of tools**—In addition to a GPIO, the BeagleBone Black has HDMI graphics output, it can act as a USB controller, and it performs various other advanced operations that are all part of system calls to an operating system. In an Arduino, you would need to hunt down and include various libraries to utilize these functions and to interact at even a low level.
- **To expand your options**—The BeagleBone Black has room for a lot more storage of programs than the Arduino. If you bring in the additional libraries to access various advanced operations, then those libraries are compiled into your program and count against your storage. You are also giving up part of your control of the bare hardware to the activities in the libraries, which could be critical depending on your application.

Also of note is the pure extra computing power of the BeagleBone Black's microprocessor over a microcontroller like the one in the Arduino. The Arduino Uno operates at 16MHz, whereas the BeagleBone Black operates at 1GHz—over 50 times the raw clock speed. There are also factors to digital circuit design that generally make microprocessors even more powerful.

So, where, you might ask, would one use an Arduino or similar microcontroller in addition to the microprocessor that is part of the BeagleBone Black? It is that bare-metal interface and the precise control of timing that makes for a great complement to the BeagleBone Black. If you have a project that requires closely coupled timing, and the uncertainty introduced by running with an operating system is too great, then you have the perfect place to offload those precision-driven capabilities to an Arduino. In fact, one of the latest Arduinos, the Arduino Yun, includes a microprocessor in addition to the microcontroller, specifically for handling some of the heavier load requirements of a network interface.

What Should Readers Get Out of This Book?

Two different types of readers can benefit from this book. There are those readers who have advanced exposure to electronics and have been programming on embedded platforms for years. They are jumping up and down at the callous simplicity of the comparisons I have made. There are other readers who have never used a computer outside of a Windows environment who are getting nervous because I've used a lot of acronyms and terms such as "compile into your program." No matter where you are on that spectrum, I want to quote Douglas Adams (author of the *Hitchhiker's Guide to the Galaxy*) by saying, "don't panic!"

By the end of this book, advanced readers should know enough about the BeagleBoard Black—and how to implement the functionality with which they are already familiar—to be dangerous. Readers who are new to all this will hopefully know enough to be even more dangerous. Therefore, we are going to start off with some basics and move forward with the goal of not leaving anyone behind. Following is a description of the chapters covered in the book:

- Chapter 2, "Introduction to the Hardware," introduces more details of the BeagleBone Black hardware and the lineage of devices that lead to the BeagleBone Black. Chapter 2 includes definitions of the specifications so that you understand the terminology.
- Chapter 3, "Getting Started," discusses how to get started with the BeagleBone Black, including how to get the basic hardware, boot up the board for the first time, and make the board do something other than run the operating system. This chapter walks you through how to connect the BeagleBone Black to the outside world. You are also introduced to the larger BeagleBone community. Rest assured, you're not alone!
- Chapter 4, "Hardware Basics," introduces you to the world of development on the BeagleBone Black. It discusses options for how to tell the device what you want it to do and the different methods used to achieve various goals. This is where we say, "Hello, World!"
- Chapter 5, "A Little Deeper into Development," explores a couple of alternate operating systems for the BeagleBone Black beyond the default operating system. One is Fedora, a well-known and well-maintained distribution of Linux. The other is Android, an operating system familiar to many smartphone and tablet users.
- Chapter 6, "Trying Other Operating Systems," delves into hooking up hardware to the GPIO pins we discussed. If you've never read a schematic before, don't worry, this chapter walks you through it as painlessly as possible. If you're an electronics pro, this chapter shows you how to implement a couple of those basic tools you already know you want to use.
- Chapter 7, "Expanding the Hardware Horizon," goes deeper into circuit development, electronics basics, and components. We expand our GPIO connections to a breadboard (see Figure 1.5) and start living a little more dangerously.

FIGURE 1.5 A breadboard with a circuit prototype. It really isn't as scary as it looks.

- Chapter 8, "Low-Level Hardware and Capes," gives you a break from developing your own hardware and talks about add-on hardware environments, known in the Beagle world as *capes,* that others have developed and how they integrate into the BeagleBone Black. It also introduces you to the idea that you can develop your own capes!
- In Chapter 9, "Interacting with Your World, Part 1: Sensors," you get exposure to how to sense what is going on in your environment. Sensors are available that correspond to all your five senses, but with capabilities far beyond those of a human. You can read that information into your project and use it!
- Chapter 10, "Remote Monitoring and Data Collection," takes the knowledge you gain from Chapter 9 on how to read sensors and shows you how to build a project to monitor some conditions in the environment around your board, and publish the data.
- The complement to Chapter 9 is Chapter 11, "Interacting with Your World, Part 2: Feedback and Actuators," which introduces actuators. Actuators are devices that let your project make things happen in the world. Servos and motors and linear actuators, oh my!
- The sensors discussed in Chapter 9 are basic sensors; this includes how to read the brightness of the environment. Chapter 12, "Computer Vision," takes this basic concept up a level and explains how to connect a camera to the BeagleBone Black. We then start combining previous lessons to create a system where your BeagleBone Black will look at your face and decide if it should open a lock.

- Chapter 13, "Sniffing Out Car Trouble," takes the show on the road, literally. Every car manufactured since 1996 has a standard computer interface. You can use the BeagleBone Black to create your very own "car computer" that tracks all the information your car is collecting. You can give that clunker some of the features of a new luxury car!
- We really get off the ground in Chapter 14, "Ground Control to Major Beagle." We start combining everything you have learned so far into a couple capstone projects. The chapter introduces something called a software-defined radio (SDR). It starts by explaining how to track local aircraft and then shows you how to receive weather data directly from satellites.
- By the time you reach Chapter 15, "Moving Forward," you'll have learned a lot of basics. You'll know enough to be either dangerous or *really* dangerous. This is a good thing. Chapter 15 gives you some ideas about other projects you can pursue and some resources for continuing your education.

I hope you make a ton of mistakes. I make mistakes all the time, and I learn a lot more from them than I do from my successes. I also want to make sure your expectations are properly set on what you *won't* get out of this book. I cover a lot on how to hook up components and what they are doing, but I only briefly delve into the "why." The components I introduce in these pages are only a small part of much larger families of components—it will be up to you to take the basics you learn here and expand this knowledge to utilize some of these other components.

Are you ready? Then let's begin.

Introduction to the Hardware

Knowing the hardware with which we will be working and the history of that hardware can be important in understanding why we do things in certain ways later on. In this chapter we will become more acquainted with the BeagleBone Black hardware and its history.

A Short Lineage of the BeagleBone Black

The BeagleBone Black is one of four boards produced by the BeagleBoard.org Foundation. It is the latest in the series of boards, and it packs a kick. The BeagleBoard.org Foundation exists for the purpose of education in open hardware and open software as well as the promotion of the ideas found in the open hardware/software community. Texas Instruments microprocessors are used at the core of all BeagleBoard devices.

The microprocessor at the heart of all of the BeagleBoard family of devices is built on the ARM architecture. ARM refers to a class of processors that utilize a specific instruction set—in this case, an instruction set developed by the British Acorn computer company. This is similar to how most of the desktop and laptop devices you used in the past were based on Intel's x86 instruction set. The ARM instruction set is, however, completely different. The ARM architecture uses a reduced instruction set code, or RISC, design. Because the architecture was developed by Acorn, this lead to the processor being called the *Acorn RISC Machine*, or ARM. This architecture means that ARM processors can do more with a little less of everything, including less power. This is what makes the ARM family of processors so popular in embedded computing.

The original BeagleBoard, shown in Figure 2.1, was introduced in 2008. It came in at a cost of $125 and was a great start to the BeagleBoard family. The original BeagleBoard sported an ARM processor clocked at 720MHz and continued to undergo revision and development up to revision D in 2012. The BeagleBoard brought to the table a set of hardware tools with all the capabilities of a laptop in a low-power, single-board profile. Of course, part of the beauty of the BeagleBoard was that it was introduced as open hardware designed to run open-source software, which meant that all the schematics of the BeagleBoard and all revisions were available to the general public, and the board design could be copied, modified, and used by anyone.

This legacy in the BeagleBoard family means that it is easy and affordable for anyone—from the home hobbyist or maker to the corporate embedded systems designer—to design and build extensions to the BeagleBone ecosystem and even replicate the architecture on a board as part of a larger embedded system design.

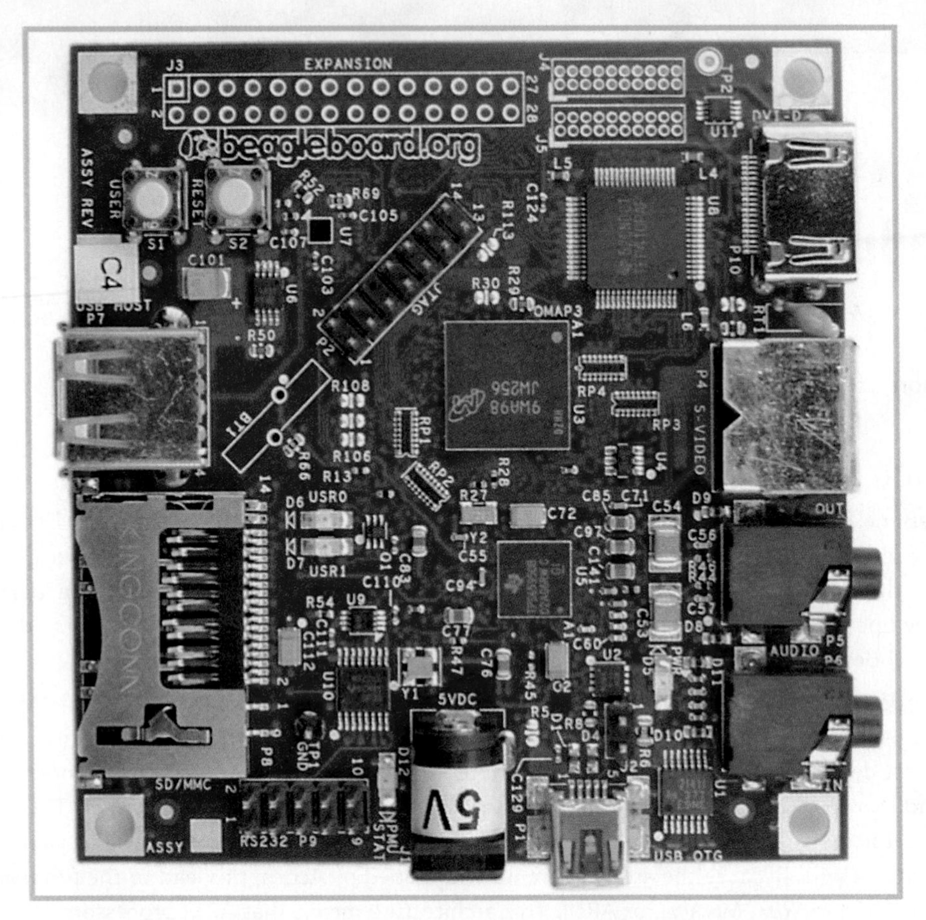

FIGURE 2.1 The original BeagleBoard (Image: BeagleBoard.org).

The next board developed was the BeagleBoard-xM, shown in Figure 2.2. It is with the BeagleBoard-xM that we start to see where the power of an open design with an active community pays off. The BeagleBoard-xM was introduced in 2010 with the cost slightly increased to $149. With the cost increase came a number of improvements over the original BeagleBoard design. Improvements included a Texas Instruments DM3730 ARM processor clocked at 1GHz rather than the 720MHz of the original BeagleBoard design. Input from the community on possible improvements to the BeagleBoard design played an important role in ensuring that the new board met the needs of application and system designers.

FIGURE 2.2 The BeagleBoard-xM (Image: BeagleBoard.org).

The next generation board was given the title BeagleBone (see Figure 2.3). This board came in at a less expensive price tag of $89 in 2011. The BeagleBone had similar performance metrics to the original BeagleBone, with a 720MHz clock on a Texas Instruments Sitara ARM processor. A major improvement to the BeagleBone over the predecessor boards was the move of the GPIO to a set of headers lining either side of the board. This footprint led to a broad, easy-to-interface ecosystem for the development of hardware extensions and even the ability to use multiple extensions by stacking up to four boards at a time. This really started to open up the floodgates in the use of the Beagle family by a larger and larger community.

FIGURE 2.3 The BeagleBone (Image: BeagleBoard.org).

One major project that really helped spark more interest in the BeagleBone was the OpenROV project (see Figure 2.4). Founded by David Lang and Eric Stackpole, the OpenROV project is an affordable, remote-controlled drone for exploring under water. I had the joy of working with the OpenROV team at the Aquarius Reef Base when it underwent its first sea trials. The original heart of the OpenROV was the BeagleBone, and as of version 2.5 of the OpenROV design, the BeagleBone Black is the main control computer.

FIGURE 2.4 The OpenROV preparing for sea trials at the Aquarius Reef Base.

The most recent incarnation of the BeagleBone is the BeagleBone Black, shown in Figure 2.5. The BeagleBone Black is the focus of the remainder of this book, and the next section of this chapter focuses on the technical details of the board hardware.

FIGURE 2.5 The BeagleBone Black (Image: BeagleBoard.org).

BeagleBone Black Hardware Specification

One of the great things about using an open-source hardware board such as the BeagleBone Black is the extensive and detailed specification of the hardware. The BeagleBone Black hardware schematics, circuit board layout, and other details are all easily available online. The details of the BeagleBone Black hardware specification are shown in Table 2.1.

TABLE 2.1 BeagleBone Black Hardware Specification

	Specification
Processor	Texas Instruments Sitara AM3358 ARM Cortex-A8 @ 1GHz
RAM	512MB DDR3
Onboard Flash	4GB eMMC
Ethernet	10/100Mb RJ45
External Storage	microSD

As previously mentioned, the BeagleBoard family has always utilized a Texas Instruments ARM Processor. For the BeagleBone Black at Revision C, the processor is a Texas Instruments Sitara AM3358 ARM Cortex-A8 family processor. The processor is clocked at 1GHz (gigahertz), or one billion cycles per second.

Despite the general idea of multitasking, a processor can only do one thing at a time per processor core. (I'm lying a little bit because certain pipelining technologies can do a couple things at a time, but I will stick with the realm of simple processor design.) The TI Sitara has a single core, so it can only perform one task at a time. The clock on the system triggers and synchronizes activities. It is the metronome for the orchestra of components on the chip, and the operating system is the conductor deciding who gets time on the hardware. Note that the clock isn't a clock counting seconds; rather, it simply switches between on and off states at a very specific cadence. The TI Sitara 3358 does, however, have what is known as a *real-time clock (RTC)* to keep track of time.

At one of the clock pulses, the microprocessor executes a machine-level instruction. This is a very low-level instruction. Rather than an instruction that in one statement opens a file, the machine-level instruction moves pieces of memory around and performs logic and mathematic operations. The hardware understands only a limited number of individual instructions. Anything more complicated is done by a combination of these instructions.

These concepts are mentioned here only as an introduction to the lower-level concepts of the brains of the BeagleBone Black. We won't use any machine-level instruction coding in this book. However, some great guides to microprocessors and machine-level instructions are available, and you can write software that works directly with the processor and forgo an operating system completely. That level of development, however, is beyond the scope of this book.

The following sections review the specifications listed in Table 2.1 individually so that you can gain a better understanding of what they mean for BeagleBone Black developers.

Processor

We touched on the processor a little bit at the beginning of this chapter. Therefore, you already know the microprocessor is an ARM device running at 1GHz—but is that all that's in the Texas Instruments chip? The microprocessor is just one of 27 subsystems built in to that one chip. The chip also contains the following components:

- Two microcontrollers that can be programmed individually and provide deterministic timing and control
- A graphics accelerator that takes some of the graphics processing workload off of the microprocessor such as rendering 3D graphics
- A separate set of controllers for mapping memory and connecting off-chip memory
- Controllers to directly drive touchscreen sensitivity and to control the liquid crystal display (LCD)
- Controllers for several communications protocols, including the following:
    - Universal Asynchronous Receiver/Transmitter (UART)
    - Universal Serial Bus (USB)
    - Inter-Integrated Circuit (I^2C)
    - Controller Area Network (CAN)
    - Hardware touchscreen readback
    - LCD/HDMI Graphics Display Control
- Ethernet controller
- Multimedia Card (MMC) control

Most of the additional subsystems perform duties for the microprocessor or allow it to interact easily with other components. A lot of stuff is packed onto that little piece of silicon. How little? The physical device, shown in Figure 2.6, is just 15mm by 15mm. A lot of what you see there is packaging. The actual silicon chip is only a fraction of that packaging. Most of the space in the package is to make room for the connectors.

All the functionality in this little chip is what makes the BeagleBone Black such a powerful platform.

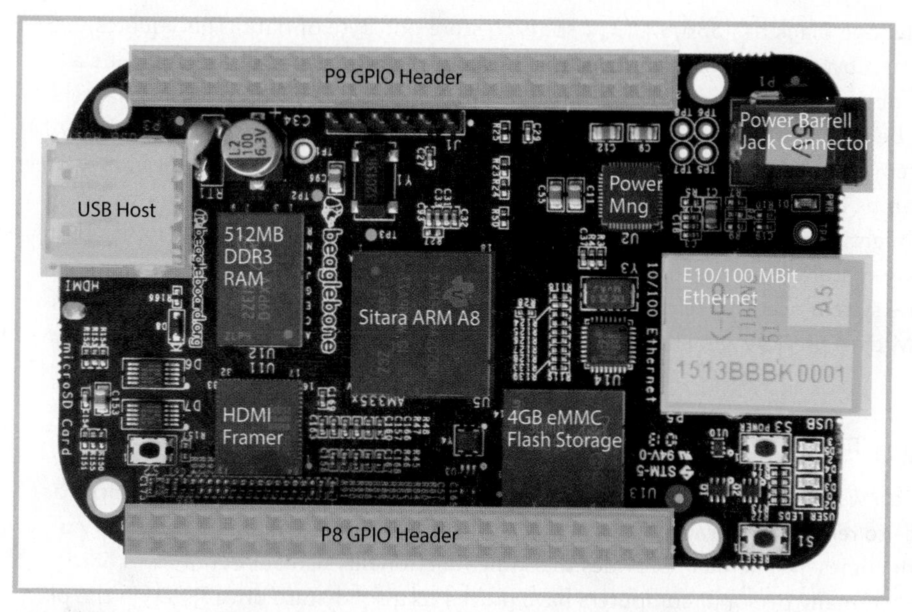

FIGURE 2.6 The Sitara AM3358 microprocessor at the heart of the BeagleBone Black.

RAM

You've probably seen the term *RAM* thrown around a lot in computer specifications, but I want to make sure you understand what RAM actually represents. RAM stands for *random access memory*. This just relates to the fact that you can access pieces of information individually in the memory rather than having to move through other data or physically move something to access the data. For example, in a hard drive, to access a piece of information, time is spent moving the discs and the read/write head to the specific location of the information. With RAM, you provide the memory system the address of the information you wish to retrieve and it is almost instantly made available to you. It is very fast, but for speed you sacrifice permanence. Once power is removed from the system, the information in most RAM devices is lost. This is known as *volatile memory*.

This volatility is acceptable, however, because the real purpose of the RAM is to store information needed by programs as they are executing. Long-term information is stored elsewhere. When a program needs a variable to store a piece of information, a space is allocated in the RAM and reading and writing to that spot occurs very quickly. The faster the better, as this affects the speed at which your program can execute!

The BeagleBone Black has 536,870,912 spots to store one byte of data. This equates to 512MB (megabytes), as 1MB is equal to 1,048,576 bytes rather than 1,000,000. (It's a binary number thing.) This is a lot of storage space, but as you will see in future chapters, there can be a lot of stuff going on at once. Many programs compete for space in the RAM. If the system starts to run out of RAM, some of the information in it that is not immediately needed can be benched, so to speak, and moved into other memory. This is known as *paging*. It might seem like a great way to boost the available memory, but the catch is that the temporary storage area is always slower than the RAM. The more RAM in your system, the less the system needs to rely on paging. The current generation BeagleBone Black has more RAM than previous generations and has a noticeable effect on how fast the system runs.

Onboard Flash and MicroSD External Storage

So where do we store information on the BeagleBone Black that won't fit into memory or that needs to remain in storage when the power is off? This is where the flash memory of the system comes in. The flash provides the same functionality as your computer's hard drive. In fact, many personal computers have moved to a solid-state drive (SSD) in the place of a traditional hard drive. These solid-state devices utilize flash memory.

The BeagleBone Black Rev C comes with 4,294,967,296 bytes of flash storage built in—that's 4GB (gigabytes). (Again, not the 4,000,000,000 you would expect, but we are dealing with binary number counting. It makes sense then. Trust me.) This is the equivalent of having a 4GB hard drive attached to your system. It isn't a ton of storage, but it's a good start for storing the operating system and some basic functionality. If you want more storage, you can add a microSD card. You may have seen these used for storing pictures in digital cameras, as extra storage in phones, or used in other devices. This capability greatly expands the possible storage space you can use.

Ethernet

The BeagleBone Black includes an Ethernet port in the hardware. This port allows you to connect your BeagleBoard Black to a network via an Ethernet cable. The maximum speed on the connection is 100Mbps (mega*bits* per second). This is a pretty good speed for almost any application you would want to run on a small device such as the BeagleBone Black. Note that in the BeagleBone Black, the physical signal interface for the Ethernet is provided by a chip that is external to the Sitara chip.

Wi-Fi networking capability can be added to the via USB as we will discuss in later chapters. Wi-Fi networking capability allows you to wirelessly connect your board to a network and can make embedding the device much easier because you don't need to ensure you have physical access to a network port.

General-Purpose Input/Output

One final (and important) specification is related to that array of connectors on the top of the BeagleBone Black. This is the general-purpose input/output, or GPIO. We will make use of the GPIO in many of the examples and projects throughout the book. In its simplest form, the GPIO provides the ability to control many of these pins however you like. Not all pins can be used in the same way, but there is enough diversity to ensure that you have the connections you need for just about any functionality you could want to add, including driving electronics and interfaces of your own design. Want to add a light to your project? Connect it through the GPIO, and you can control when the light is on or off. Have a communications protocol you want to use that's not already supported by the board? Allocate a set of GPIO pins and you can program your own interface. These pins also provide access to many of the built-in protocols as well as power and ground.

The GPIO is physically divided into two clusters: P8 and P9. These are the physical names of the headers. The individual pins are numbered, but they may have different names in the logical programming realm. I will try to keep it clear throughout the book which pin I am referring to, both physically and logically.

Each header has 46 pin locations where a wire can be inserted. Many of the pins have multiple modes of use, but we will get into that in a later chapter. Table 2.2 provides a list of the pinouts.

TABLE 2.2 BeagleBone Black Expansion Headers

P9				P8			
Ground	1	2	Ground	Ground	1	2	Ground
+3.3V Power	3	4	+3.3V Power	GPIO	3	4	GPIO
+5V Input Power	5	6	+5V Input Power	GPIO	5	6	GPIO
+5V System Power	7	8	+5V System Power	GPIO	7	8	GPIO
+5V Logic Level	9	10	System Reset	GPIO	9	10	GPIO
GPIO	11	12	GPIO	GPIO	11	12	GPIO
GPIO	13	14	GPIO	GPIO	13	14	GPIO
GPIO	15	16	GPIO	GPIO	15	16	GPIO
GPIO	17	18	GPIO	GPIO	17	18	GPIO
GPIO	19	20	GPIO	GPIO	19	20	GPIO
GPIO	21	22	GPIO	GPIO	21	22	GPIO
GPIO	23	24	GPIO	GPIO	23	24	GPIO
GPIO	25	26	GPIO	GPIO	25	26	GPIO
GPIO	27	28	GPIO	GPIO	27	28	GPIO

P9				P8			
GPIO	29	30	GPIO	GPIO	29	30	GPIO
GPIO	31	32	ADC Ref Voltage	GPIO	31	32	GPIO
Analog Input	33	34	Analog Ground	GPIO	33	34	GPIO
Analog Input	35	36	Analog Input	GPIO	35	36	GPIO
Analog Input	37	38	Analog Input	GPIO	37	38	GPIO
Analog Input	39	40	Analog Input	GPIO	39	40	GPIO
GPIO	41	42	GPIO	GPIO	41	42	GPIO
Ground	43	44	Ground	GPIO	43	44	GPIO
Ground	45	46	Ground	GPIO	45	46	GPIO

All these GPIO pins can have multiple mode and uses. Depending on what you have connected to the system, many pins may not be available. In Chapter 4, "A Little Deeper into Development," I discuss different programming languages and how to access these pins, but I'll go over a couple of the conventions now. The libraries for BeagleBone Black GPIO access from Adafruit Industries provide a very simple way to select which pin you are programming. It is a string that is built as follows:

```
string := P<header>_<pin>
header := 8 or 9
pin := 1…46
```

So, for example, if you want to access GPIO pin 35 on header P8, the string is this:

```
P8_35
```

Easy enough, right? This is similar, as you will see in Chapter 4, in the BoneScript language, but it is a variable naming convention rather than a string-based convention. However, accessing the pins through the operating system or using C/C++ libraries is slightly more complex. For most of the book, I will stick with either BoneScript or Python but will make sure to be clear on any complexities and provide reference information when the usage isn't straightforward.

Getting Started

To get started with the BeagleBone Black, you will want to obtain the board itself and some of the basic parts. The board is available from a number of resources. The beagleboard.org website provides some great places to buy the board (see http://beagleboard.org/black). Of course, I have my favorite suppliers, which are both a major part of the Maker community:

- **SparkFun Electronics (http://sparkfun.com)**—Boulder, Colorado–based SparkFun Electronics was founded on the idea of open source and open hardware. The company has an extensive catalog of electronics parts and components and consistently supports the community. SparkFun also has excellent tutorials and active forums where you can always find help on your projects. If you live in the Boulder area, be sure to watch for the in-person classes and other events the company sponsors!
- **Adafruit (http://adafruit.com)**—Based in New York City, Adafruit was founded by Limor "Ladyada" Fried, a true celebrity in the Maker community. The company is another great source for all the components and parts you might need for your projects and for taking the steps to move your projects beyond the introductory ones in this book. Adafruit also has an extensive tutorial section on its website.

When I am shopping for parts, I am often torn between these two suppliers. They are both excellent companies that have provided great support to me and countless others. Determining which to use, however, shouldn't be as much of a struggle as I might make it seem. The two companies work well together, participate in events together, and are both friendly. They represent the best of the community because they aren't there to compete against each other; they support the community and make sure we have what we need to get our projects off the ground. If you have questions on electronics basics or more complicated techniques, you can generally find a tutorial on one of the two sites or find willing support in the forums and via email.

In each chapter, I will outline the parts you will need for that chapter's project. We will take a practice run with this chapter and look at getting your BeagleBone Black set up and running. First, note some of the parts you'll need:

- BeagleBone Black
- USB cable (USB A to USB Mini B)
- +5 volt DC power supply (at least 1,000 milliamps)
- Ethernet cable

The USB cable should come with the BeagleBone Black; if it does not, make sure you inform your supplier. If a friend gave you the board, then just ask nicely. To get started, you also need a way to communicate with the board. In this chapter I'm going to discuss how to connect the BeagleBone Black to another computer and also how to connect remotely via Ethernet. Another option is to connect a monitor, keyboard, and mouse directly.

Setting Up and Saying "Hello, World!"

The big moment has arrived, and it is time to power up your BeagleBone Black and start working! We start with a direct computer connection via USB. This is a simple step. Simply plug the USB cable into the BeagleBone Black and the other end into a USB port on the computer.

As soon as the board is connected to your computer, you should see the lights on the board come to life. Four lights should start blinking on the board. These are four "user" lights, labeled USR0, USR1, USR2, and USR3, as shown in Figure 3.1. There is also a power light labeled PWR. The power light should stay constantly lit. The user lights will blink with different activities. At the default boot on a clean board, you will find the user lights configured as follows:

- **USR0**—This light blinks in a heartbeat pattern: two quick flashes, a pause, and then repeat.
- **USR1**—This light is configured to blink on activity from the microSD card. Because the board isn't plugged in a microSD card yet, you shouldn't see any activity on this light yet.
- **USR2**—This light flashes on CPU activity.
- **USR3**—This light flashes when the built-in flash memory is accessed. The default operating system should be installed in this embedded Multi-Media Card, or eMMC, memory, so you should see activity as the board is accessing the built-in, default operating system and file system.

FIGURE 3.1 The BeagleBone Black user lights, power light, and USB port highlighted.

Next, you're going to have to install the drivers necessary for your computer to talk to the BeagleBone Black. On a Windows 7 laptop, you just plug the board in via USB to allow a drive to mount from the board that contains driver files. Use these files directly so that you don't have to look for them. You also have the option of downloading the drivers from the BeagleBoard organization website (http://beagleboard.org/getting-started). Drivers are available for Windows 32- and 64-bit environments, OS X, and Linux. You should refer to the specific setup instructions for your machine.

Now that you have your BeagleBone Black up and operating and have the drivers installed, what can you do? Your BeagleBone Black is already running a web server! You can use the Chrome or Firefox web browser to navigate to the board's web server at http://192.168.7.2.

NOTE

Web Browser Warning

The web server is not compatible with Microsoft Internet Explorer. Just stick with Chrome- or Firefox-compatible browsers. You really shouldn't use Internet Explorer anyway. For anything. Take this as a public service message. You can get Chrome or Firefox at the following links:

http://www.google.com/chrome/browser

http://www.mozilla.org/en-US/firefox/new

Navigate to the BeagleBone Black website at http://192.168.7.2. Your browser should present a very colorful and active website, and you should see something like the banner shown in Figure 3.2.

FIGURE 3.2 Banner from the BeagleBone built-in website letting you know everything is working just fine.

This banner gives you information about your BeagleBone Black. First, the banner is green and has a check mark. That must mean everything is good, right? Also, the banner tells you that the board is connected. These are all good indications that all is right with your BeagleBone Black's world. Table 3.1 explains the other information that is supplied.

TABLE 3.1 Default Website Banner Information

Listed Information	Description
BeagleBone Black	You bought a BeagleBone Black, right? This is a good sign that you bought what you thought you were buying.
Rev 000C	This tells you that, in this case, version 000 of the revision C BeagleBone Black hardware is running.
S/N 2314BBBK0577	This is the serial number of the specific BeagleBone Black.
Running BoneScript 0.2.4	The board runs BoneScript version 0.2.4. BoneScript is a version of JavaScript for the BeagleBone environments.
At 192.168.7.2	This is the IP address of the board on the virtualized network over the USB connection. It should match the address you typed into the address bar of the browser.

Congratulations! You've now successfully powered up, connected to, and communicated with your board! That was easy, wasn't it? Let's make a couple changes and use one of the user lights we discussed before to make it flash. About halfway down the page is a section titled "Cloud9 IDE" (IDE stands for *integrated development environment*). Click the header, and the Cloud9 IDE will launch in a new browser window or tab (see Figure 3.3). This is a powerful IDE running directly on the BeagleBone Black through a web interface.

So, what is an IDE? In short, an IDE is used as an all-in-one place where you can write software directly on the BeagleBone Black. It includes an editor, a way to execute code, and many other useful features.

When a person is just starting out with a new programming language, there is a tradition that the first program they write simply displays "Hello, World!" in some manner that fits into the environment. In many languages, this is accomplished by simply printing the message, whereas in some Windows environments, an alert message is displayed. That tradition has been extended into the hardware world with a program that makes a light blink once a second.

In our case, we are approaching a board for the first time and trying out a language for the first time, so why don't we try both displaying a message and blinking a light? Follow these steps to create a new file, write the code to accomplish our task, execute the code, and get blinking:

1. In the main window of the Cloud9 environment you'll find a + button. Click this button and select New File. This will open a blank text file where you can enter code. If there are other tabs open, you can close them. Feel free to peruse any "getting started" information on those pages.
2. Enter the code shown in Listing 3.1 into the document.
3. Save the file on the board. In this case, the file's name is blink.js.
4. In the environment, click the Run button.

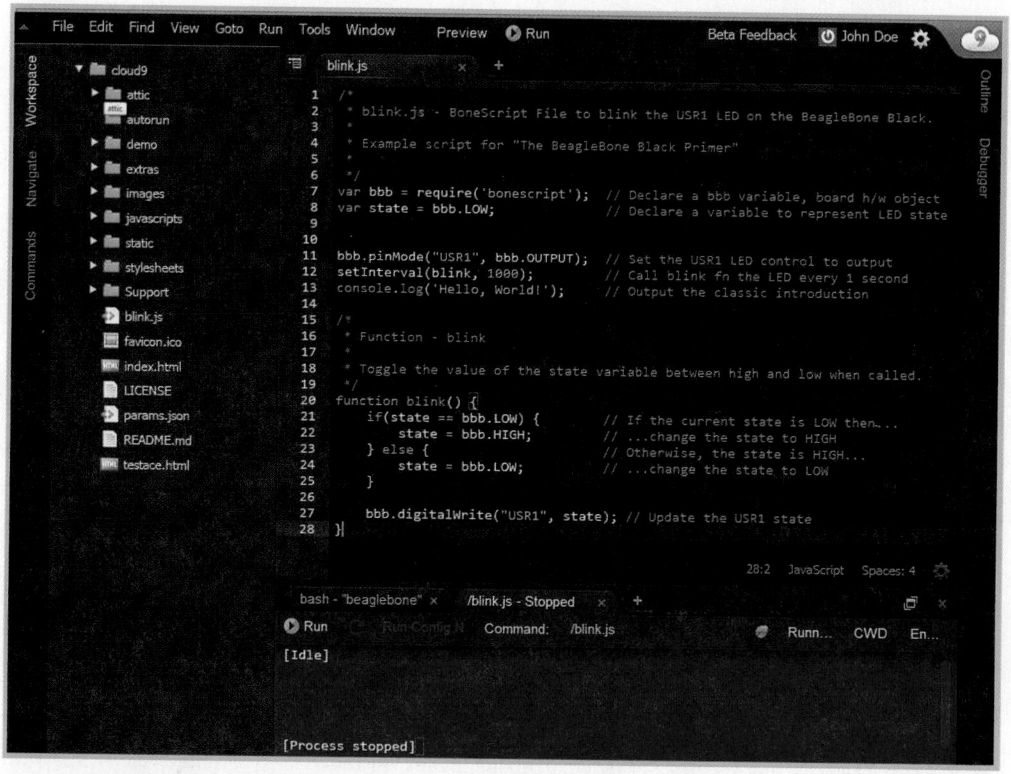

FIGURE 3.3 The Cloud9 IDE running on the BeagleBone Black.

LISTING 3.1 blink.js

```
1:  /*
2:   * blink.js - BoneScript File to blink the USR1 LED on the BeagleBone Black.
3:   *
4:   * Example script for "The BeagleBone Black Primer"
5:   *
6:   */
7:  var bbb = require('bonescript');  // Declare a bbb variable, board h/w object
8:  var state = bbb.LOW;               // Declare a variable to represent LED state
9:
10:
11: bbb.pinMode('USR1', bbb.OUTPUT);  // Set the USR1 LED control to output
12: setInterval(blink, 1000);          // Call blink fn the LED every 1 second
13: console.log('Hello, World!');     // Output the classic introduction
14:
15: /*
```

```
16:   * Function - blink
17:   *
18:   * Toggle the value of the state variable between high and low when called.
19:   */
20: function blink() {
21:      if(state == bbb.LOW) {           // If the current state is LOW then...
22:           state = bbb.HIGH;           // ...change the state to HIGH
23:      } else {                         // Otherwise, the state is HIGH...
24:           state = bbb.LOW;            // ...change the state to LOW
25:      }
26:
27:      bbb.digitalWrite('USR1', state); // Update the USR1 state
28: }
```

It will take a couple of seconds, but the code will start executing. You should see a light just next to the heartbeat light blinking on for a full second and then off for a second. Success!

Let's step through the code you just blindly put into the environment and executed. Glad you trust me! The source starts with these six lines of code:

```
1:   /*
2:    * blink.js - BoneScript File to blink the USR1 LED on the BeagleBone Black.
3:    *
4:    * Example script for "The BeagleBone Black Primer"
5:    *
6:    */
```

These lines look fairly readable to a human and not like source code. That's because this code is what's called a *comment*. A comment starts with `/*` and ends with the `*/` and includes everything in between. The extra asterisks at the beginning of the other lines are just to make things look good. There is another way to signify comments in BoneScript/JavaScript, and that is using `//`. These are used to describe what is occurring on a line of code. Everything from the `//` to the end of the line is a comment. Comments are not executed or even seen for execution. You will see a couple of different styles of commenting in different languages throughout the book.

Line 7 accesses a shared library of source code, called `bonescript`, that is provided to you as part of the environment:

```
7:  var bbb = require('bonescript');  // Declare a bbb variable, board h/w object
```

This code accomplishes many tasks behind the scene that you don't need to worry about for now. Access to the library is assigned to variable `bbb`. This means that we can use the variable `bbb` to access resources in that special library, as you will see on the following lines:

```
8:  var state = bbb.LOW;              // Declare a variable to represent LED state
```

Line 8 declares another variable called `state`. We are going to use `state` to track whether we set our signal for the USR1 light to `HIGH` or `LOW`. When the state is set to `HIGH`, the voltage on the electronics attached to that light is set to +5V, and it is set to 0V for `LOW`. When the voltage is set `HIGH` at +5V, the electrical potential on the light is increased, which means the light can do work. What happens when a light can do work? It lights up!

Something important to remember here is that setting the `state` variable to `HIGH` or `LOW` doesn't actually change the power supplied. We do that using a function called `digitalWrite`, which is a part of the `bonescript` library we can now access through the `bbb` variable. More on that function later. Now we hit a line that does something with the electricity on the board:

```
11: bbb.pinMode('USR1', bbb.OUTPUT);   // Set the USR1 LED control to output
```

With this line, we are calling a function called `pinMode`, which is part of the `bonescript` library, and using another `bonescript` library constant called `OUTPUT`. This means we are configuring the USR1 pin to output the voltage rather than sensing a voltage from somewhere else in a circuit. In total, this line says, "Take the pin attached to the light USR1 and get it ready to output, please."

The next line utilizes a function called `setInterval` to run the meat of the program:

```
12: setInterval(blink, 1000);          // Call blink fn the LED every 1 second
```

This line of code tells the system to execute the function `blink` once every second. Line 13 has nothing to do with blinking our light. This is a simple statement that prints our classic first-time program announcement out to a console:

```
13: console.log('Hello, World!');      // Output the classic introduction
```

In the Cloud9 IDE environment, you will see this printed on a lower tab labeled "/blink.js - Running," as shown in Figure 3.4.

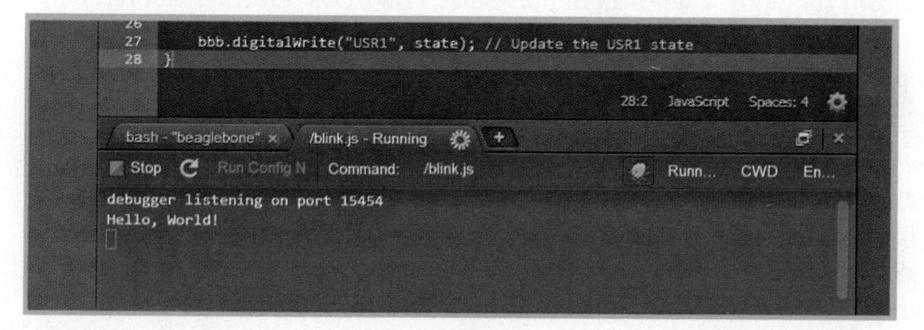

FIGURE 3.4 The "Hello, World!" statement written to the console log.

The final lines define a function called `blink`. This function simply checks the status of the `state` variable and changes it to the opposite state. This function is called once every second by the `setInterval` function. The real meat of the function is on line 27. The call to `digitalWrite` makes the actual change to the hardware to change the status of the physical circuit attached to the USR1 light:

```
27:     bbb.digitalWrite('USR1', state); // Update the USR1 state
```

That is all the code required to use BoneScript to blink a light and print a message to the console! It is important to remember that BoneScript is defined only by the `bonescript` library. The underlying syntax and structure is just JavaScript, a scripting language used in many places on the Web. This means that you can use JavaScript tutorial and reference resources to help you understand or to get any clarification.

For simple examples throughout the book, I will stick to BoneScript just to make it easy. For more complex code and functionality, I use other programming languages such as C/C++ and Python. I will comment the code to help with readability if you are not familiar with those languages; however, I highly encourage you to seek out other resources to learn those languages in depth because that is not the focus of this book. The next chapter will delve into some more complex development with BoneScript and the Cloud9 IDE to enable your own explorations. It will also introduce you to the basics of programming with other languages.

Connecting to Ethernet

Thus far, we have talked to the BeagleBone Black through a USB connection to a computer. This is all fine and well, but the power of the BeagleBone Black is that it's a standalone computer capable of working on its own. Our next step is to cut the cord from our computer and connect the BeagleBone Black to a network.

In accomplishing this, we can drop the USB cable from our setup and exchange it for an Ethernet cable. We also need to power our board. Ethernet, unlike USB, does not provide power in normal configurations. There's an option called Power over Ethernet, abbreviated PoE, but this is not a normal network configuration, so we will assume you need a separate power supply. I purchased the power supply I am using from SparkFun. It has a 5V output and can provide up to 1A of current.

Most home networks use a system called Dynamic Host Configuration Protocol (DHCP). In this configuration, a component on a network is assigned an address on the network automatically. To make this easy on ourselves, we know when the BeagleBone Black is connected via USB that it has an address of 192.168.7.2. We can use this to our advantage and connect to both Ethernet and USB at the same time and see what address our board is assigned for the Ethernet connection. So, with your board already connected to the USB, plug your Ethernet cable into the board and into your network.

When you connect the Ethernet cable, you should see the lights on the Ethernet port of your board light up. This means you've connected! Now, to see the IP address that has been assigned, we are going to break out to a new piece of software and connect via Secure Shell (SSH). We are about to delve into the world of the Linux command line.

There are many ways to connect via SSH. If your computer runs Linux or OS X, getting to a terminal is as easy as opening a Terminal session. The commands to use are the same as you would see working on the command line of a Linux or OS X machine. Following is an example of connecting via SSH from the Linux command line first. The process is the same for OS X. We will get to Windows in a moment.

From the Terminal, execute the following command:

```
[brian@mercury-fedora-vm ]$ ssh root@192.168.7.2
```

With this command, you ask the system to use the `ssh` command to connect as `root` to the computer at `192.168.7.2`, which we know is our BeagleBone Black's USB connection. When you execute this command and you have all the connections set right, you are presented with the following prompt:

```
The authenticity of host '192.168.7.2 (192.168.7.2)' can't be established.
ECDSA key fingerprint is c0:81:1a:f4:58:b9:51:15:00:df:ee:71:c4:d9:fd:54.
Are you sure you want to continue connecting (yes/no)?
```

This prompt is associated with security (referring to the first *S* in SSH). Your computer has never connected to this host via SSH before, and it wants to validate that this is the computer you mean to connect with. What does this buy you? It allows you to be sure that the computer you are connecting to in the future is the one you intended to, with no hacker interference. You should accept this by entering **yes**. The `ssh` program will let you know that it has accepted the security key and that it is added to the list of known servers.

```
The authenticity of host '192.168.7.2 (192.168.7.2)' can't be established.
ECDSA key fingerprint is c0:81:1a:f4:58:b9:51:15:00:df:ee:71:c4:d9:fd:54.
Are you sure you want to continue connecting (yes/no)? yes
Warning: Permanently added '192.168.7.2' (ECDSA) to the list of known hosts.
```

Now, you request to log in as the root user. The root account is a powerful account and should be password protected by default, but the BeagleBone Black has a blank root password by default. We will change this later before we set the board up to be on a network doing a job.

In the Windows environment, I recommend PuTTY for SSH connections. It is easy to find with a Google search, and installation is a breeze. When you start the application, you are presented with the configuration window shown in Figure 3.5. Notice in the hostname that I've entered the USB assigned address of your BeagleBone Black, 192.168.7.2. Just below the Host Name text entry, you select the connection type, which is SSH in this case. Once you've entered these settings, click the Open button.

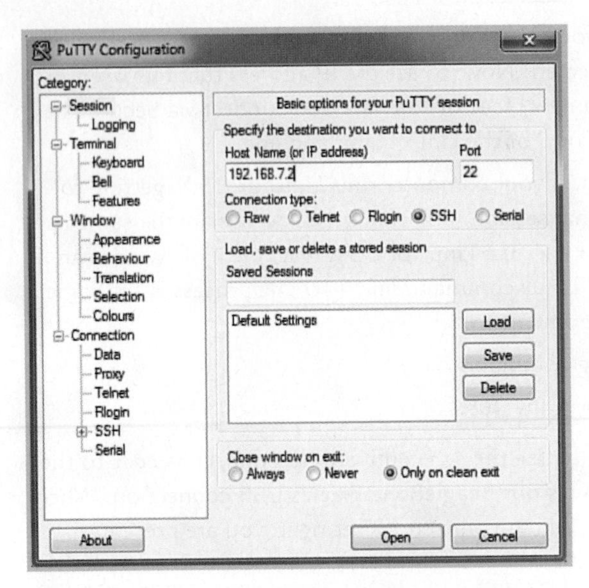

FIGURE 3.5 The PuTTY Configuration window.

Another window will pop up that looks a lot like the information you saw the first time you tried to connect in the Linux terminal, and it serves the same function (see Figure 3.6). Click the Yes button to accept the security key and continue by logging on.

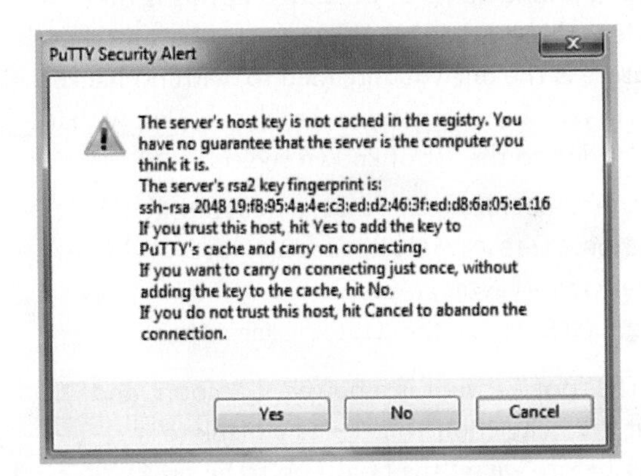

FIGURE 3.6 The PuTTY Security Alert window.

From here on, regardless of the operating system or Terminal application you are using, the output will be the same. That is because what you'll see now is actually on the BeagleBone Black.

From now on, if you log in with SSH via the USB default connection, you will not see the prompt for the security key. The session will now present you with the following command prompt:

```
root@beaglebone:#
```

This is the default prompt for the default user. If you are familiar with Linux or a similar operating system, then you'll know you're in a Bash shell. The information provided by the prompt can be very useful and even customized. The information in Table 3.2 is presented in the default prompt.

TABLE 3.2 Default Prompt Information

Prompt Information	Description
`root`	The information in this first block tells us the username for the shell. In our case, we logged in as root.
`beaglebone`	The hostname, on the network, we are logged in to.
	This represents the directory we're currently working in on the file tree. In this case, the tilde is shorthand for the user's home directory.

Now, we are going to enter our second command. This command, called `ifconfig`, is used to report the current network status of the system. Let's go ahead and enter it and then see the response:

```
root@beaglebone:# ifconfig
eth0      Link encap:Ethernet  HWaddr 7c:66:9d:58:bd:41
          inet addr:192.168.1.161  Bcast:192.168.1.255  Mask:255.255.255.0
          inet6 addr: fe80::7e66:9dff:fe58:bd41/64 Scope:Link
          UP BROADCAST RUNNING MULTICAST  MTU:1500  Metric:1
          RX packets:4059 errors:0 dropped:2 overruns:0 frame:0
          TX packets:147 errors:0 dropped:0 overruns:0 carrier:0
          collisions:0 txqueuelen:1000
          RX bytes:616100 (601.6 KiB)  TX bytes:18322 (17.8 KiB)
          Interrupt:40

lo        Link encap:Local Loopback
          inet addr:127.0.0.1  Mask:255.0.0.0
          inet6 addr: ::1/128 Scope:Host
          UP LOOPBACK RUNNING  MTU:65536  Metric:1
          RX packets:0 errors:0 dropped:0 overruns:0 frame:0
          TX packets:0 errors:0 dropped:0 overruns:0 carrier:0
          collisions:0 txqueuelen:0
          RX bytes:0 (0.0 B)  TX bytes:0 (0.0 B)

usb0      Link encap:Ethernet  HWaddr e6:8c:89:9a:b6:c8
          inet addr:192.168.7.2  Bcast:192.168.7.3  Mask:255.255.255.252
          inet6 addr: fe80::e48c:89ff:fe9a:b6c8/64 Scope:Link
          UP BROADCAST RUNNING MULTICAST  MTU:1500  Metric:1
```

```
RX packets:1717 errors:0 dropped:0 overruns:0 frame:0
TX packets:136 errors:0 dropped:0 overruns:0 carrier:0
collisions:0 txqueuelen:1000
RX bytes:200409 (195.7 KiB)  TX bytes:31059 (30.3 KiB)
```

The results provided tell us about three different network adapters represented on the system: `eth0`, `lo`, and `usb0`. We can ignore `lo` for now; it's the local loopback connection. The two of interest to us are `eth0` and `usb0`. The default USB connection is `usb0`. There are a lot of fields here, but the field we are interested in is labeled `inet addr`. Here is the `usb0` interface information again, with that field highlighted in bold:

```
usb0      Link encap:Ethernet  HWaddr e6:8c:89:9a:b6:c8
          inet addr:192.168.7.2  Bcast:192.168.7.3  Mask:255.255.255.252
          inet6 addr: fe80::e48c:89ff:fe9a:b6c8/64 Scope:Link
          UP BROADCAST RUNNING MULTICAST  MTU:1500  Metric:1
          RX packets:1717 errors:0 dropped:0 overruns:0 frame:0
          TX packets:136 errors:0 dropped:0 overruns:0 carrier:0
          collisions:0 txqueuelen:1000
          RX bytes:200409 (195.7 KiB)  TX bytes:31059 (30.3 KiB)
```

The value associated with this address should look familiar. It is the same address we used to access the website and to log in to the board. Now, what we are looking for is the address that has been given to the board via DHCP. The Ethernet port is represented by interface `eth0`, and by looking at its `inet addr` field, we know that, in this case, the DHCP has assigned the board an address of `192.168.1.161`, as shown here:

```
eth0      Link encap:Ethernet  HWaddr 7c:66:9d:58:bd:41
          inet addr:192.168.1.161  Bcast:192.168.1.255  Mask:255.255.255.0
          inet6 addr: fe80::7e66:9dff:fe58:bd41/64 Scope:Link
          UP BROADCAST RUNNING MULTICAST  MTU:1500  Metric:1
          RX packets:4059 errors:0 dropped:2 overruns:0 frame:0
          TX packets:147 errors:0 dropped:0 overruns:0 carrier:0
          collisions:0 txqueuelen:1000
          RX bytes:616100 (601.6 KiB)  TX bytes:18322 (17.8 KiB)
          Interrupt:40
```

Unless you have changed the IP address space in your network, and if you have I trust that you know what you are doing with your network, your board will have an IP Address in one of the public IP Address spaces, either 192.168.x.x or 10.0.x.x.

Now, with the board connected to Ethernet and assigned an address on the network, you can unplug the USB connection and plug in the +5V power adapter. You've now put your BeagleBone Black on the network, independent of the computer you were using before. You are ready to be an active member of your home's network ecosystem!

Let's check the website connection via your network connection. Using your browser, navigate to the address provided earlier for `eth0`. In the case of my network, that's 192.168.1.161, as you can see in Figure 3.7.

FIGURE 3.7 Banner from the BeagleBone built-in website letting you know everything is working just fine—this time, via the Ethernet connection.

Now, let's log in via SSH to the `eth0` connection. It is going to look familiar:

```
[brian@mercury-fedora-vm ]$ ssh root@192.168.1.161
The authenticity of host '192.168.1.161 (192.168.1.161)' can't be established.
ECDSA key fingerprint is c0:81:1a:f4:58:b9:51:15:00:df:ee:71:c4:d9:fd:54.
Are you sure you want to continue connecting (yes/no)? yes
Warning: Permanently added '192.168.1.161' (ECDSA) to the list of known hosts.
Debian GNU/Linux 7

BeagleBoard.org BeagleBone Debian Image 2014-04-23

Support/FAQ: http://elinux.org/Beagleboard:BeagleBoneBlack_Debian
Last login: Fri Jul 18 15:06:44 2014 from mercury-win.local
root@beaglebone:#
```

As you can see, it is the same process for logging in as before, but with a different address. This example is from a Linux machine, but the process is identical as the one we used previously for PuTTY, but with a different address.

Something interesting to note is that if you run the `ifconfig` command, the `usb0` adapter is still available. That's because we haven't actually changed anything in the operating system configuration; we just used a different connection. The USB option is still there waiting for us to connect, and it will be unless we turn it off in the operating system. We will get into more detail about the underlying operating system and some of the options we have for alternate operating systems on the BeagleBone Black. In our next chapter, we delve into the hardware and discuss some basics of electronics that will be necessary for many of the remaining chapters.

4

Hardware Basics

This chapter is a short tutorial on electronics and electricity. If you know little to nothing about electronics, you will know enough to be dangerous by the end of the chapter. It takes entire textbooks to cover some electronics-related topics. If you want to delve deeper into electronics, I encourage you to look at some other great books specifically related to that subject.

For this chapter, you will need a BeagleBone Black board (the examples are built on Rev C).

Electronics Basics: Voltage, Current, Power, and Resistance

In the previous chapter, we connected our BeagleBone Black to our computer via USB, and then through a +5 V DC power adapter. How did the board receive power over the USB connection? Simple, it is part of the USB specification. Table 4.1 shows the power specification that USB must meet to supply power.

TABLE 4.1 USB Power Specifications

Specification	Maximum Current	Minimum Voltage	Nominal Voltage	Maximum Voltage	Power Supplied
USB 1.0	100mA	+4.75V	+5V	+5.25V	0.75W
USB 2.0	500mA	+4.75V	+5V	+5.25V	2.5W
USB 3.0	900mA	+4.45V	+5V	+5.25V	4.5W

Table 4.1 introduces one of the basic laws of electricity. You'll notice that the values and ranges for voltage, current, and power vary and that as more current is drawn, the overall amount of power supplied is higher. This is because voltage, current, and power are all related and interconnected measures of electricity. Let's look at what each of these values means at a high level.

- **Volt**—The *volt* is a measure of electrical potential. Do you remember basic physics? There are two types of energy: potential and kinetic. If something is sitting on the floor, it has zero energy. If you pick it up and hold it five feet off the ground, you've now given it potential energy. If you let go of that object, it drops to the floor, and when it hits the ground, it releases that potential energy into the ground as kinetic energy. Electrical potential is similar and gives you an idea of how much energy is available. The symbol we use for the volt is "V."

- **Ampere**—The *ampere*, or *amp*, is a measure of how fast the electricity flows through a conductor. The higher the ampere value, the more electricity flows past the point you are measuring at a time. If you want to do work with the electricity, you are going to slow the flow down and need to make sure you have enough current to accomplish the task at hand. The symbol for an amp is "a." Just to be confusing, whereas the amp is the measure of current, the symbol for current itself is "I." There are reasons for this, I promise.
- **Watts**—If you imagine the amount of electricity flowing past a point with a certain amount of potential, you are now measuring the power in the system, measured in *Watts*. The symbol for the Watt is "W." The Watt is the unit of measure, but in equations, power is represented by "P."

These three terms connect through a very simple equation:

$$P = I \times V$$

Isn't the universe wonderful that such an important relationship is so simple? It becomes incredibly simple to see what happens to two other variables if one variable is varied. The best way to think of this is in the triangle shown in Figure 4.1.

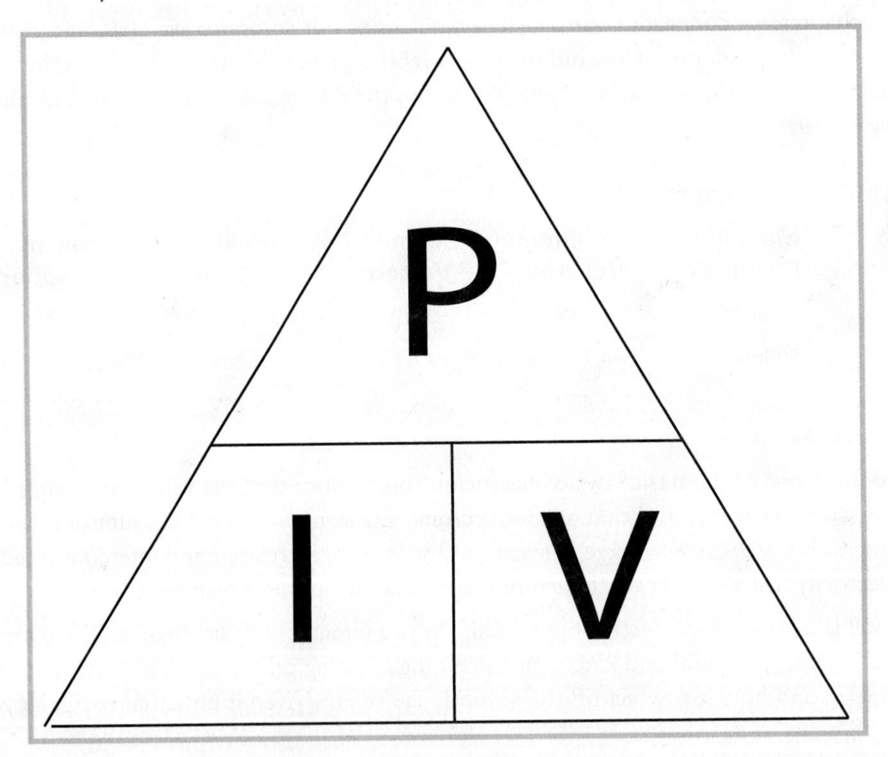

FIGURE 4.1 The triangle of power.

The way to use this triangle is simple. If you need power, you multiply the current and the voltage to get power. If you want to get current from the power and the voltage, you divide the power by the voltage. And to get the voltage? Just divide the power by the current. Piece of cake!

You actually see this relationship at work around you. Have you ever plugged a vacuum in, turned it on, and seen all the lights attached to that circuit in your house dim? It takes a lot of power to turn a vacuum motor! We know the voltage of household power is about 120V, and that is roughly fixed, give or take a little wiggle room. We know that the vacuum needs a certain amount of power to operate. What is left to change? The current! The vacuum is now using more current from the circuit, which means there is less current for the lights! Now, because there is less current for the lights, and the voltage is staying at 120 volts, what does that mean there is less of in the lights? Power! With less power, the lights dim! What happens if you add another vacuum and need more than the 15 amps available on the circuit? You're changing a fuse or flipping a circuit breaker.

So how much power does a BeagleBone Black actually use? Table 4.2 shows some of the reference data for the board for power consumption. This is highly dependent on what hardware is attached and what activity is going on at the same time. We know the board is running on +5V, and the BeagleBone Black System Reference Manual provides reference currents. Utilizing the simple relationship, you can also determine how much power is utilized. Table 4.2 is specific to the BeagleBone Black Rev C hardware and can vary slightly based on your model and revision.

TABLE 4.2 BeagleBone Black Power Utilization

Mode	Current (mA)	Power (W)
Boot Idle	210	1.05
Kernel Boot	460	2.3
Kernel Idle	350	1.75
Webpage Load	430	2.15

The power running through a circuit makes *work* happen. This concept of work introduces another important concept, *resistance*. Whenever you put something in the flow of anything and ask it to do work, that flow will get slowed down as it passes through. A river can run faster when it doesn't meet a waterwheel. A horse can run faster when it doesn't have to pull a wagon. So, how can we measure resistance? Because power, current, and voltage are all related, and power is a measure of the work that can be done, it follows that resistance can fit in the equation as well. The equation is, in fact, simple and no more difficult than the power equation. In fact, you can also represent resistance with a simple relational graphic, shown in Figure 4.2. The unit to measure resistance is the Ohm and it uses the Greek symbol Ω.

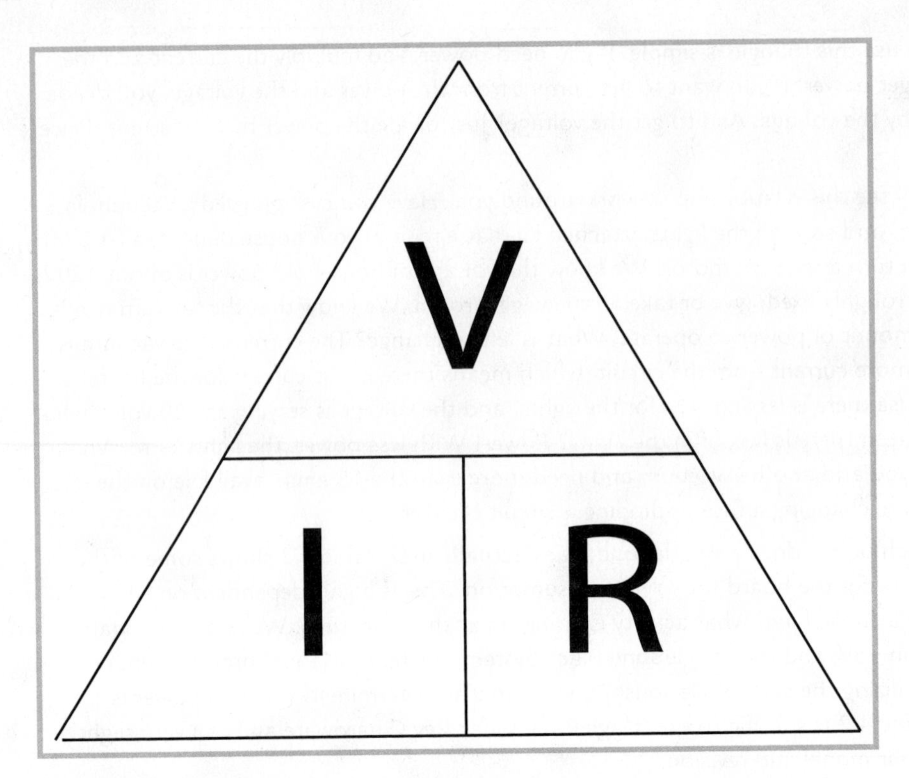

FIGURE 4.2 The triangle of resistance.

You can use this triangle for resistance exactly as the power triangle:

$$R = \frac{V}{I}$$

$$I = \frac{V}{R}$$

$$V = I \times R$$

Now, we know voltage and current are two players in both the power formula and the resistance formula, so there must be a way for all these to work together. They must all be related! Figure 4.3 shows how they all work together in a well-known form.

With this graphic, the relationship becomes obvious, and finding the value of one parameter from a combination of the others is trivial. The wheel divides into four quadrants. The equations in the upper left provide an answer as power, the upper right as voltage, lower right as resistance, and lower left as current.

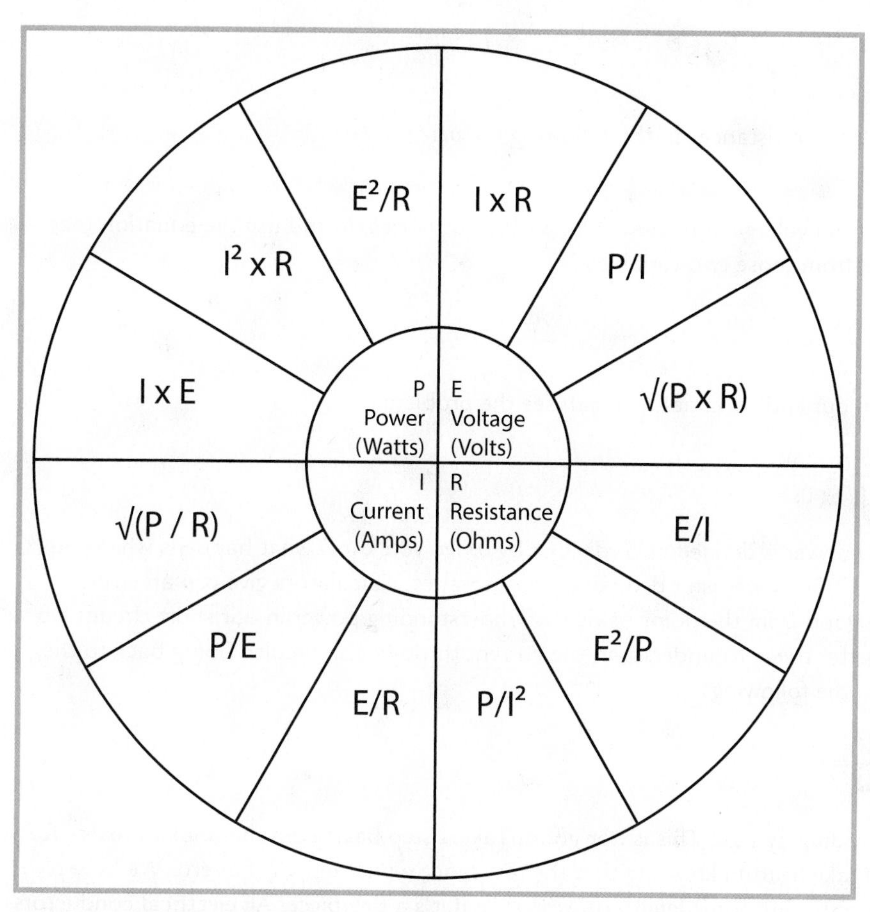

FIGURE 4.3 Circle of electrical power.

The Short Circuit

One important concept needs to be addressed now. It is a term often used but not well understood by newcomers. The equations discussed in the previous section all involve *power*, which is work the circuit can do, and *resistance*, which is the slowing down of that flow to do the work. But what happens if you connect a power source from beginning to end but don't have it do any work in between? You get a short circuit!

So, what does a short circuit mean from the standpoint of the previous equations? The best way to look at it is by trying to measure the power or the current flow in that situation. Let's take the USB power we talked about in Table 4.1. We know we get a voltage of roughly +5V from the USB supply and that it can provide up to 500mA in the case of the USB 2.0 specification. What happens if we connect it up with nothing to do work? Well, let's start by seeing how much power it will use.

If we know that

$$V = 5V$$

then there isn't any resistance, so the following is true:

$$R = 0\Omega$$

To get power from voltage and resistance, we look at our circle and use the equation that gives us power from those two variables:

$$P = \frac{V^2}{R}$$

Substituting in our known values, you can see the problem:

$$P = \frac{V^2}{R} = \frac{5^2}{0} = \frac{25V}{0\Omega} = ???W$$

Therefore, our power is defined as 5^2 divided by 0. Do you know what happens when you divide by zero? The answer isn't defined in mathematics. Calculators give you an error. That can't be good from the point of view of understanding power in our short circuit. Do we fare any better trying to understand the current through our circuit? Going back to the wheel, we find the following:

$$I = \frac{V}{R} = \frac{5V}{0\Omega} = ???A$$

We are still dividing by zero. This is not good. Take a step back from the brink of reality for a second and take heart in knowing that the resistance won't actually be zero. We have to connect the circuit with some length of wire, even if it's a tiny piece. All electrical conductors have some resistance of their own, albeit very tiny in most cases. What if we substitute that tiny value for our zero? When I measured on my meter, with the leads, I come up with a value of 1.3 Ω, but when I calibrate just the leads, I come up with a value of[el] 1.3 Ω, as shown in Figure 4.4. This is just a limitation of my meter, because I know the resistance is nonzero. The best we can glean from this is that the value is likely no more than 0.09 Ω. It is probably far less, but we will go with that value. Therefore, you have the following calculation:

$$P = \frac{V^2}{R} = \frac{5^2V}{0.09\Omega} = \frac{25V}{0.09\Omega} =\sim 278W$$

FIGURE 4.4 Resistance of a one-inch piece of wire, with and without leads.

Looking at the table for USB power, we see that 278W is a little more power than the 2.5W the standard can supply. How do we fare with the current in the circuit?

$$I = \frac{V}{R} = \frac{5V}{0.09\Omega} =\sim 55.5A$$

Your household power is likely limited to 200 amps, so this represents about one-quarter of the power that can flow into your house at any moment. What really happens in a short circuit is that all that energy potential quickly transforms to kinetic energy. It is released until something stops the flow, such as a fuse, a circuit breaker, or the components involved melting. Strange things start happening when you play near the realm of undefined mathematics, so when I say we need to avoid shorting a circuit, you now know what I mean.

The Resistor

What is the most basic way to regulate the flow and increase the resistance? Something called a *resistor*. A resistor has a simple function in a circuit, but is used everywhere and comes in all shapes and sizes. In our flow analogies, this is like putting a speed bump in a circuit. The cost is that the resistor uses a little of the work the circuit is capable of doing and gets rid of that work as heat. The gain is that we now don't just have voltage running through a circuit at whatever current it wants to flow. We have taken our first steps to controlling a circuit!

A resistor, in U.S.-based schematics, is a squiggly line. European standards show the symbol as an empty, rectangular box. Figure 4.5 shows a picture of a resistor and both of the schematic representations. Note that the remainder of this book uses the U.S. symbol.

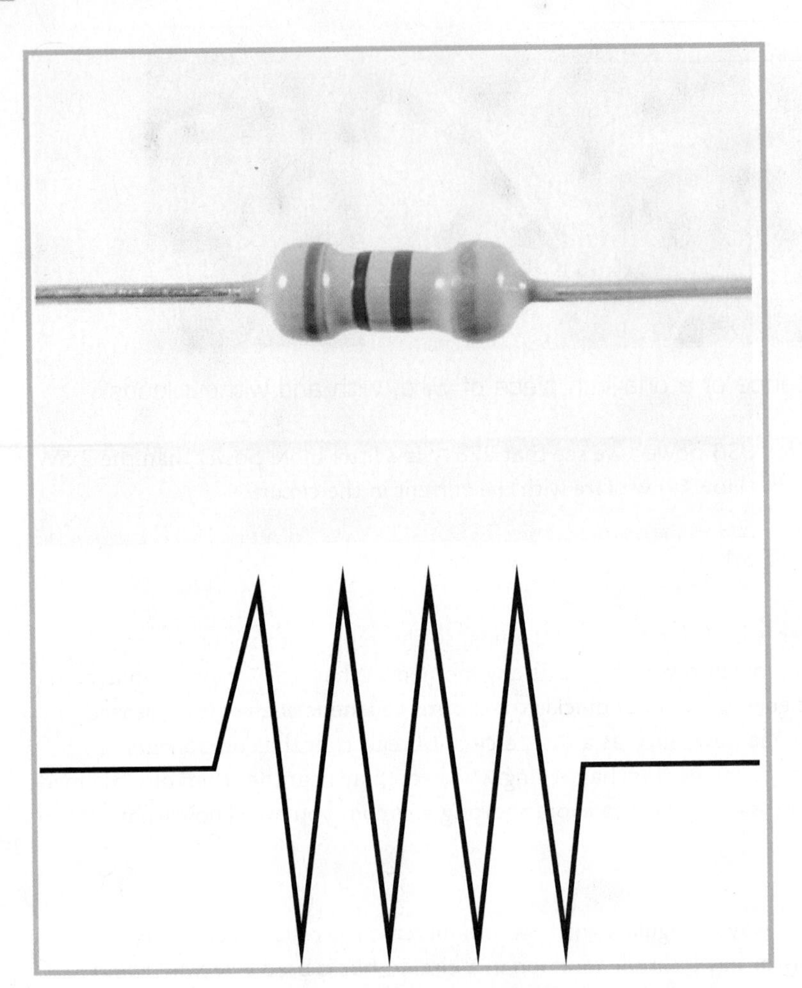

FIGURE 4.5 A resistor and the schematic representation.

Identifying a resistor's value is easy, but not as easy as reading a number off the packaging. Resistors have colored bands wrapped around the body. The color of these bands indicates the resistance of the component. The Table 4.3 provides a sense of how to read these bands.

TABLE 4.3 Resistor Band Translation Guide

Color	Band A	Band B	Band C (Multiplier)	Band D (Tolerance)
Black	0	0	1	
Brown	1	1	10	± 1.00%
Red	2	2	100	± 2.00%

Color	Band A	Band B	Band C (Multiplier)	Band D (Tolerance)
Orange	3	3	1000	
Yellow	4	4	10000	
Green	5	5	100000	± 0.50%
Blue	6	6	1000000	± 0.25%
Violet	7	7	10000000	± 0.10%
Grey	8	8		± 0.05%
White	9	9		
Gold			0.1	± 5.00%
Silver			0.01	± 10.00%

Here are a couple notes about this chart and the value convention for resistors. First, you will generally see four bands, but sometimes you will see five bands. The five-band system is used in military specification for parts, and it indicates the reliability of the component. Feel free to consult the Department of Defense handbook *Resistors, Selection and Use Of* (MIL-HDBK-199), but for this book we will stay away from that level of requirement.

Second, most resistors you see will have a gold or silver tolerance band, indicating a tolerance of ±5% and ±10%, respectively. The higher tolerances are not generally necessary, and we will not need that level of precision in the resistors we use in this book.

Just for practice, let's look at Figure 4.6 as practice for how to decode resistors.

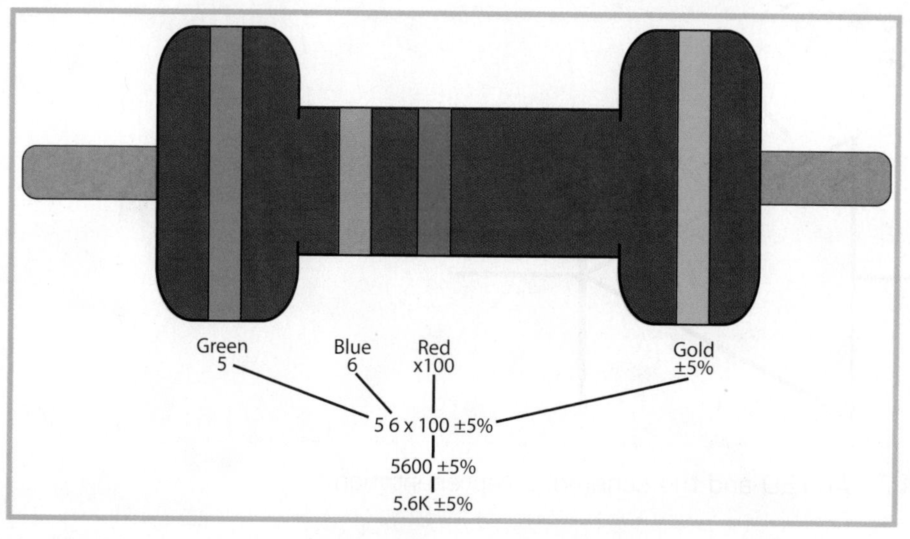

FIGURE 4.6 Resistor decoding example.

A reasonable first question is, "How do I know which end to start reading the bands?" If you look at Table 4.3, you will find that gold and silver are not used for the first band. So, when you go to decode your color bands, the gold or silver band goes on the right. For the first two bands you concatenate the numbers together, so for the green and blue bands in Figure 4.6, you get 5 and 6. Therefore, the final resistor value is a multiple of 56. Now, reading the third band, you can see that a red band means you multiply 56 by 100, giving you 5,600. You have a resistor with 5,600 Ω, or more often stated as 5.6 kΩ (read as 5.6 kilo-ohms). Now, we cannot forget the tolerance. The resistor in Figure 4.6 has a gold band and thus a tolerance of ±5%. This means that the value of the resistor is anywhere from 5% less (or 5,320 Ω) to 5% more (or 5,880 Ω). These may seem like big swings from the 5.6 kΩ you were aiming for earlier, but it is close enough for the projects you are working on. In fact, you will most likely see all 5% resistors in this book because my personal stockpile is composed of 5% resistors.

Diodes and LEDs

A *diode* controls the direction current can flow. This is useful for making sure a current doesn't go in the wrong direction and accidentally cook part of your circuit. Diodes can, however, get much more complicated. They can be designed to only let current through as long as a voltage is above a certain level or other useful properties. One type of diode we will encounter repeatedly in this book is the light-emitting diode (LED). It is a descriptive but not creative name. Remember that light we controlled on the BeagleBone Black in Chapter 3, "Getting Started"? That is actually an LED. In an LED, when the appropriate voltage is applied, the diode emits light. Figure 4.7 shows the schematic diagram for an LED as well as a picture of an LED.

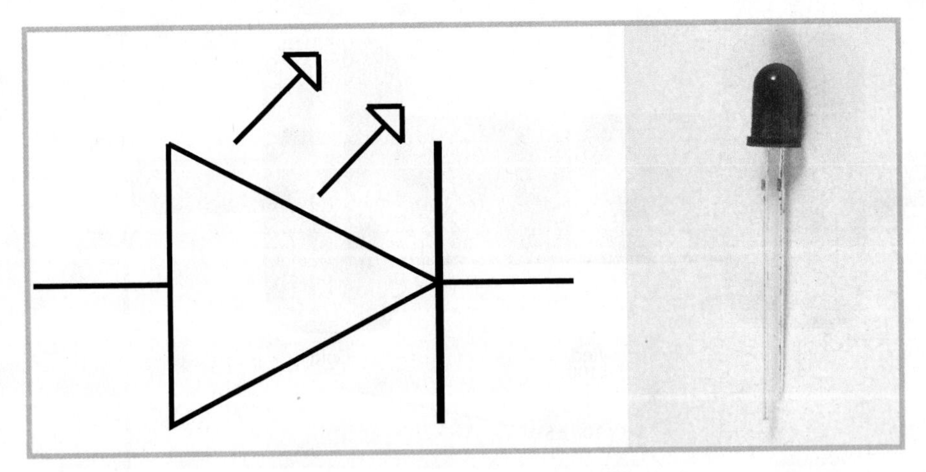

FIGURE 4.7 An LED and the schematic representation.

The LED has a specific current it works best with, and you want to make sure you keep the current flowing through the circuit near that current or you will damage the LED over time. So, how do you limit the current flow? With a resistor! First, you need to know how much resistance you will need. This is a simple formula based on the resistance triangle. An LED will change the voltage in operation, so you need to take that into account. Other than that, you will see it is the same formula you used before!

$$R = \frac{V}{I} = \frac{V_{sup} - V_{LED}}{I_{LED}}$$

Simple! Let's assume you are using an LED where you know the voltage drop is 2V and the maximum current is 30mA. Let's also assume you are driving this LED from one of the ports on the BeagleBone Black, which outputs at 3.3V. It is here that you will run into a problem. The GPIO ports on the board provide a solid 3.3V but can only supply 4mA, so that will have to be the current restriction. In a later chapter you learn about a component called a *transistor* that enables you to control more than the 4mA of current that is restricted at the GPIO pins, thus allowing for a number of high-power options under the BeagleBone Black control.

Substituting all of that, you can get the resistance you need to limit the current properly:

$$R = \frac{V}{I} = \frac{V_{sup} - V_{LED}}{I_{LED}} = \frac{3.3V - 2.0V}{4mA} = \frac{1V}{0.004A} = 250\Omega$$

Now, here's something to think about: Remember all those ± values for tolerances? The voltage varies based on the color of the LED. The variance runs from 1.8V on the low end for red to 2.4V on the high end for clear white. Also, I'm working with ±±5% resistors. That leaves some uncertainty. What is the resistor necessary for the lowest voltage possible? It is as follows:

$$R = \frac{V}{I} = \frac{V_{sup} - V_{LED}}{I_{LED}} = \frac{3.3V - 1.8V}{4mA} = \frac{1.5V}{0.004A} = 375\Omega$$

How about the highest voltage in the pack? It is as follows:

$$R = \frac{V}{I} = \frac{V_{sup} - V_{LED}}{I_{LED}} = \frac{3.3V - 2.4V}{4mA} = \frac{0.9V}{0.004A} = 225\Omega$$

Clearly, you want to go with the higher end. If you choose a smaller generic resistor and decide to swap out the LED without recalculating, you will pull too much current for the GPIO pin. You need about 375 Ω but, remember, you are using ±5% resistors. That means the resistor could be as low as 356 Ω. Yikes, you are still at risk of burning out the GPIO. So, what you want to do is make sure you choose a resistor safely outside of tolerance becoming a factor. As an added protection, because this isn't a precision circuit, you also want to protect against someone accidentally using a ±10% resistor. So, pick a resistor that is higher than the following:

$$R = 375\Omega * (1 + 10\%) = 412.5\Omega$$

In my personal supply of resistors, I don't have a 412.5 Ω resistor, but I do have some 470 Ω resistors. In this case, I would just go with the 470 Ω and be happy.

Build an LED Circuit

Why don't you go ahead and dive into building you own circuit! Here is what you want to model with the board: a simple railroad crossing blinking light pair, such as one with a red light on either side of the sign and alternate flashing. When one is on, the other is off, and vice versa. You are going to add two LEDs to the BeagleBone Black through something called a *breadboard*. A breadboard allows you to add components into a circuit without permanently wiring them to test out your design. Figure 4.8 shows a breadboard.

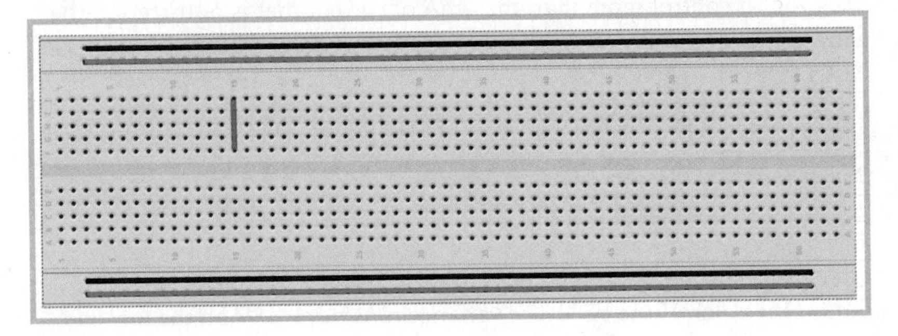

FIGURE 4.8 A breadboard.

This is a typical breadboard, although you will often also see half-length boards as well. The rows with the red line covering them are intended to carry power. You would connect these to a general power source. The rows covered by a black line are intended to be connected to ground. No, not the actual floor. *Ground* is where all current wants to flow and, generally enough, represents the 0V reference in a circuit. If you are using a battery, this is usually the negative end of the battery.

The red line of one side is all connected. This means that as you wire your power supply into one spot on that row, it is connected to all the spots on that row. Same for the ground. The power row and ground row on one side of the board are not connected to each other.

The green line highlights one column of the breadboard. Each column on one side of the board is connected together. For example, the points at 15 F through J are all connected together, but they are not connected to F A through E.

We will get more into breadboards and some more complex breadboarding lessons in Chapter 7, "Expanding the Hardware Horizon."

Now you need to design the circuit. You will take your first step into a schematic! You will use a program similar to the blink listing in Chapter 3, but will use two ports on the GPIO rather than the USR1. You will also use a ground pin on the GPIO for the electrical ground, or 0V reference. Select GPIO1_28 at P9-12 and GPIO1_16 at P9-15 for source ports. These are the ports you will set high and low in the program. So, with that knowledge, you can begin drawing up the schematic, shown in Figure 4.9.

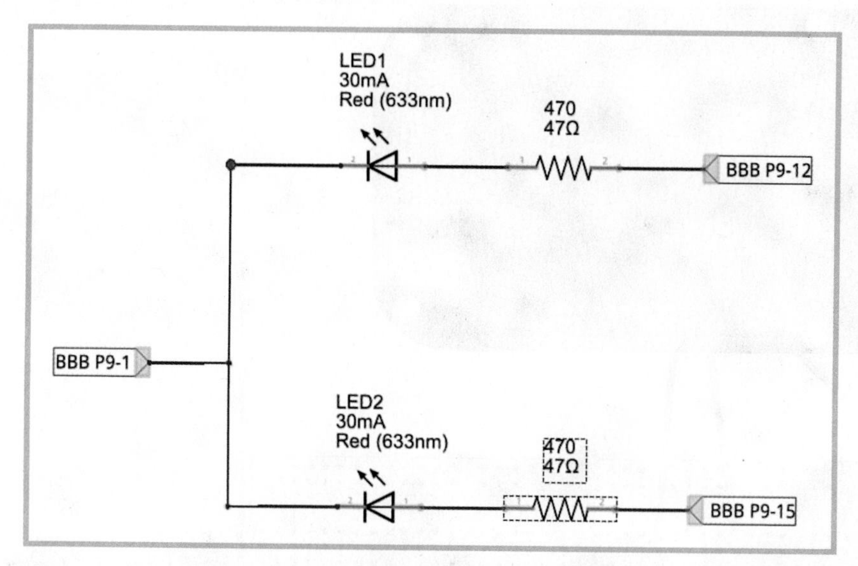

FIGURE 4.9 Basic schematic for a two-LED system connected to the BeagleBone Black.

To the right side of the schematic, you can see two entry points into the circuit. These are reference connections to the two selected pins on the BeagleBone Black. Therefore, the current will flow from those pins through a resistor at 470 Ω, just as we had discussed previously. Next, the current flows through the LED. It is important to note that the LED is known as "polarity sensitive," whereas a resistor is not. This means the orientation of a resistor doesn't matter, but the LED wants one side connected to the higher voltage and the other end toward ground. Generally, the lead on an LED has an end that is slightly longer and connects to the higher voltage end of the circuit, whereas the shorter lead connects to the lower voltage, or ground. Also, you can use any columns you want on the board; it doesn't have to be the columns I used in the breadboard illustration. There is nothing special about those columns.

Finally, you have another reference to a pin on the BeagleBone Black, P9-1. We haven't talked about that pin yet. That is because that pin is a general ground spot on the board. There is not a configuration for the pin in the software. You connect back to this ground so that you are working on a common ground from the perspective of the board and the circuit. Also note that because as ground is ground, the circuit can actually combine at the point where it connects to ground.

Now that you understand the circuit, we can move to the breadboard. You use short lengths of wires called *jumper wires* for this kind of breadboard connection. A visualization of the breadboarded circuit is provided in Figure 4.10

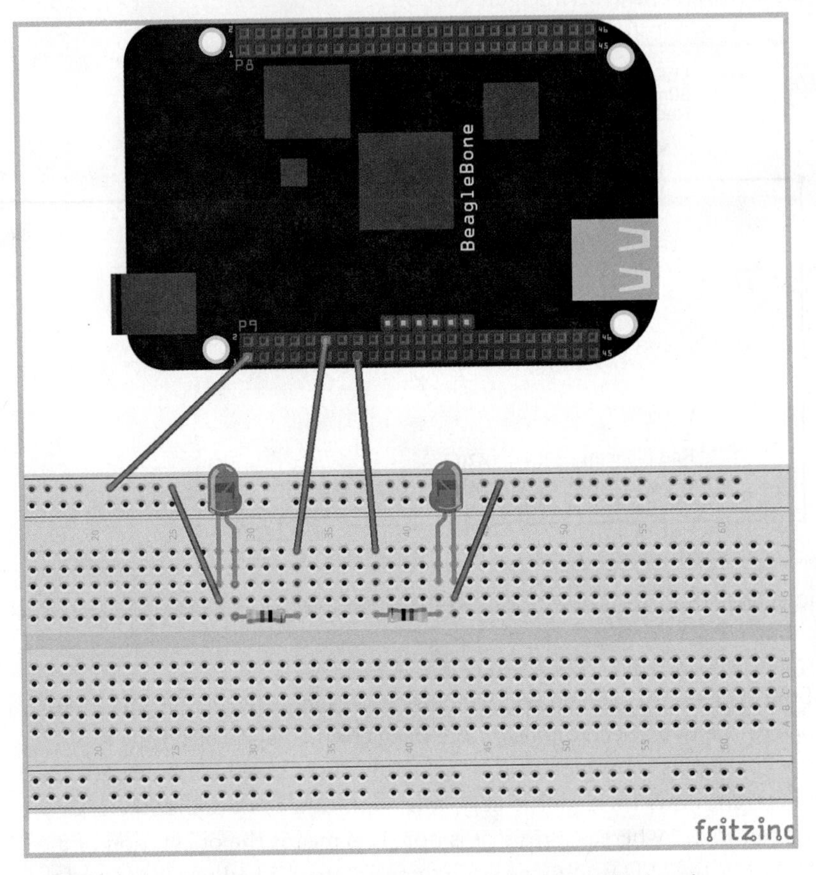

FIGURE 4.10 An illustration of the full, connected circuit.

If you go ahead and wire up your BeagleBone Black like this, right now, you won't see much happen. We need to write the software to control the port. Listing 4.1 shows the code that will operate the circuit the way you would expect.

LISTING 4.1 Circuit Operation

```
1:  /*
2:    * rr_crossing_blinkers.js - BoneScript File to blink LEDs attached to
3:    *                                GPIO P9_12 and P9_15.
4:    *
```

```
5:    * Example script for "The BeagleBone Black Primer"
6:    *
7:    */
8:   var bbb = require('bonescript');  // Declare a bbb variable, board h/w object
9:   var state1 = bbb.LOW;         // Declare a variable to represent GPIO P9_12 state
10:  var state2 = bbb.LOW;         // Declare a variable to represent GPIO P9_15 state
11:
12:
13:  bbb.pinMode("P9_12", bbb.OUTPUT); // Set the GPIO P9_12 control to output
14:  bbb.pinMode("P9_15", bbb.OUTPUT); // Set the GPIO P9_15 control to output
15:  setInterval(blink, 1000);          // Alternate blinking LEDs every 1 second
16:
17:  /*
18:   * Function - blink
19:   *
20:   * Toggle the value of the state variable between high and low when called.
21:   */
22:  function blink() {
23:      if(state1 == bbb.LOW) {        // If P9_12 is LOW...
24:          state1 = bbb.HIGH;         // ... set P9_12 to HIGH
25:          state2 = bbb.LOW;          // ... set P9_15 to LOW
26:      } else {                        // Otherwise...
27:          state1 = bbb.LOW;          // ... set P9_12 to LOW
28:          state2 = bbb.HIGH;         // ... set P9_15 to HIGH
29:      }
30:
31:      bbb.digitalWrite("P9_12", state1); // Update the GPIO P9_12 state
32:      bbb.digitalWrite("P9_15", state2); // Update the GPIO P9_15 state
33:  }
```

This code should look familiar to you because it is based directly off Listing 3.1, our basic blink program from Chapter 3. The difference here is that you are now talking to two ports, so you need to keep track of the state of those ports. You declare those two variables in lines 9 and 10. You can name the variables anything you want, but it is best to keep them named something consistent. For the sake of pure optimization and good coding practice, there are a few changes I would make, but I primarily want to make sure the functionality is clear here. We will get a little more into coding standards and styles in Chapter 5, "A Little Deeper into Development."

You can now copy this code directly into the Cloud9 IDE, just like you did in Chapter 3, and execute the program. If you've connected your breadboard up, you should see the alternating blinking. Success! My operating circuit is shown in Figure 4.11.

FIGURE 4.11 My circuit, up and running!

It looks like they are switching over instantly, doesn't it? As a prelude to the next chapter, where we will talk more about programming language options, let's look at lines 31 and 32 of our code for a second:

```
31:        bbb.digitalWrite("P9_12", state1); // Update the GPIO P9_12 state
32:        bbb.digitalWrite("P9_15", state2); // Update the GPIO P9_15 state
```

These two lines change the state of the pins on the BeagleBone Black. To our eyes, they look like they happen at exactly the same time, yet we know the lines execute one at a time. This is a good chance to pull out a classic electronics debugging tool, the *oscilloscope*. An oscilloscope simply shows us a plot of an electrical property over time. In this case, let's look at the voltage coming from our pins while the circuit is running. Figure 4.12 shows a zoomed-out view of these voltage changing.

As you can see, everything looks like you would expect. When the P9-12 pin goes high, the P9-15 pin goes low, and vice versa. This all looks like it is happening at exactly the same time. Why don't you zoom in closer? In Figure 4.12, each vertical dotted line represents 400ms. That looks about right, every 2.5 divisions is 1 second, the time we set in `setInterval()`. What happens when you zoom way down to 1ms per division?

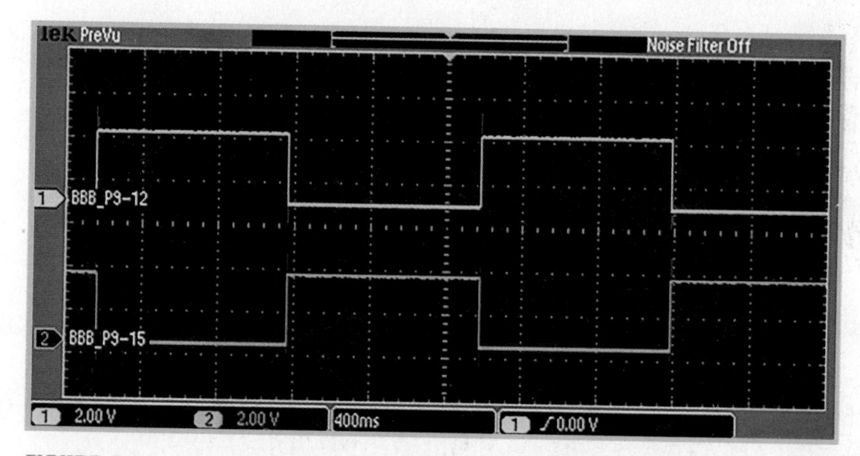

FIGURE 4.12 The oscilloscope view of our circuit running, zoomed to a high level.

As you can see in Figure 4.13, you are certainly not looking at two events happening simultaneously. They occur approximately 1ms apart. What this allows us to see is that things don't happen as fast as we may perceive.

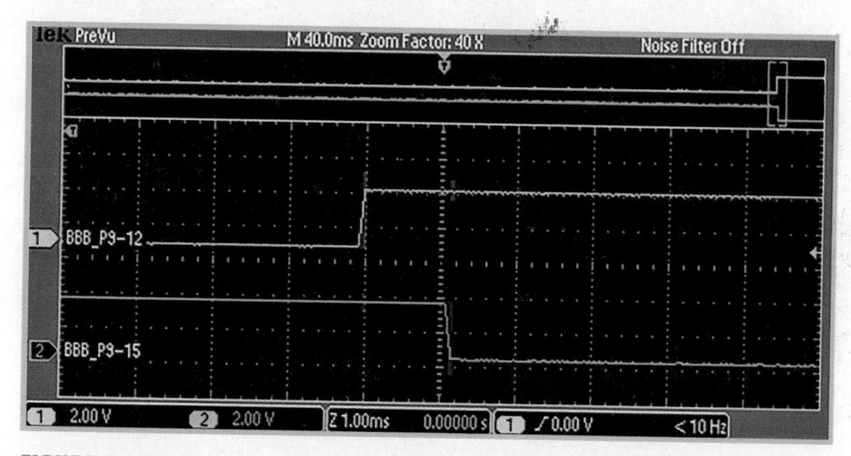

FIGURE 4.13 The same waveform capture, zoomed way down.

Are there faster ways to execute this code? The next chapter discusses programming on a slightly more advanced level and introduces other programming languages. We wil also take a closer look at the timing for executing code.

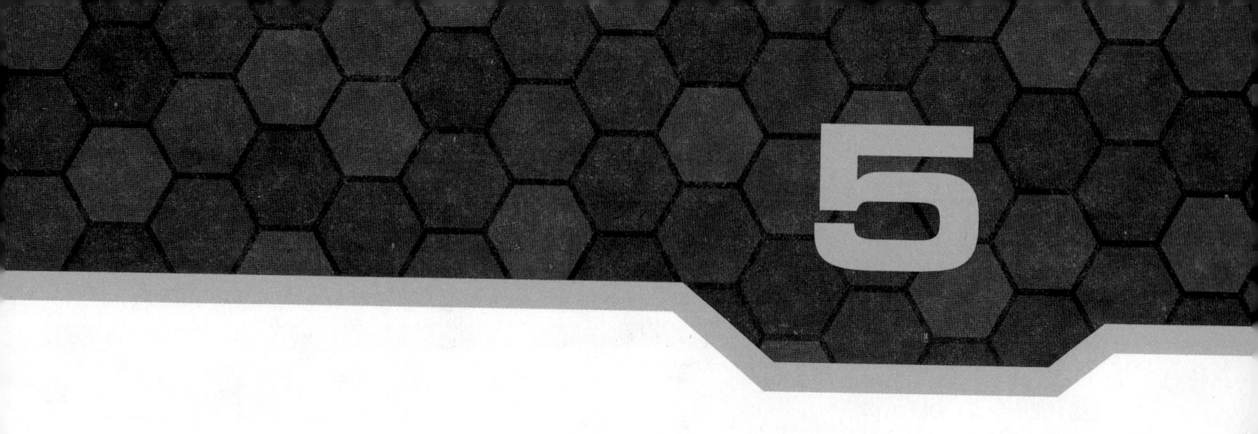

A Little Deeper into Development

So far, we have written a couple basic programs together—one of which you can find copies of on the Internet and is a basic example of how to blink a Light Emitting Diode (LED) that is built in to the BeagleBone Black. The other is an introduction to creating a circuit that uses components to alternately blink two LEDs. At the end of the last chapter, you discovered that the time between the actions for the two LEDs is 1ms.

You learned before that the Central Processing Unit (CPU) clock on the BeagleBone Black, which regulates how fast actions happen in the CPU, runs at 1GHz. That is one billion cycles per second. To get the amount of time between cycles, you simply take the inverse, and you end up with 0.000001ms between each cycle.

In this chapter, we investigate what is going on in those million clock cycles, how we can make our programs more efficient, and how to ensure that code is easily readable by anyone it is shared with. We also look at other programming languages in this chapter.

Interpreted Code

We've already talked a little about machine code. Machine code makes up the lowest-level instructions that actually tell the processor what to do. If you have two instructions to execute one after the other, and they each only take one clock cycle, then they will execute in 0.000001ms each or 0.000002 ms total.

All programs we write must, eventually, undergo transformation into a set of machine-level instructions. There are two different ways to accomplish this task: interpreted code and compiled code. In either method, most single lines of source code need to transform into several lines of machine code.

The software we have written so far is BoneScript. BoneScript is a version of JavaScript and an interpreted language. What does it mean to be an interpreted language? The best way to understand this is to step through the process:

1. Write your software and save it to a file. (We've done this twice now.)
2. Click the Run button in the Cloud9 Integrated Development Environment (IDE). The interpreter is automatically called on your file, as well as a great deal of utility work to enable debugging.

3. The interpreter looks at the file to figure out what to do with the code presented. It ignores all the comments you have included and any blank lines. Those are to make it easier for people to read the code and provide a better understanding of what is going on in the code. At this stage, the source code from Listing 4.1 looks, to the interpreter, like Listing 5.1.

LISTING 5.1 blink.js

```
1:  var bbb = require('bonescript');
2:  var state1 = bbb.LOW;
3:  var state2 = bbb.LOW;
4:  bbb.pinMode("P9_12", bbb.OUTPUT);
5:  bbb.pinMode("P9_15", bbb.OUTPUT);
6:  setInterval(blink, 1000);
7:  function blink() {
8:       if(state1 == bbb.LOW) {
9:            state1 = bbb.HIGH;
10:           state2 = bbb.LOW;
11:      } else {
12:           state1 = bbb.LOW;
13:           state2 = bbb.HIGH;
14:      }
15:      bbb.digitalWrite("P9_12", state1);
16:      bbb.digitalWrite("P9_15", state2);
17: }
```

With the insight we already have into this code, it is clear what is happening, but it would make a little less sense to someone looking at the code for the first time. It is not unreadable, but takes work to understand.

4. The interpreter begins to execute the code, including all the code from outside libraries via the `require()` function call. As the interpreter begins executing the call tree, it decides what sections of code to execute and in what order. Each line of code is turned into as many lines of machine code as is required to execute on the fly.

The interpreter executes on the same processor as the script and translates each line of code into the equivalent machine code sections and then executes those lines. We are starting to get a little insight into how the execution of a line of code can take a million clock cycles.

Another consideration is that we are running inside of the Cloud9 IDE debugger. This means that the interpreter and debugger communicate about the position of execution through the code—yet more things happening that can take up clock cycles.

The code is also running in an environment where a number of other things are running. The main BeagleBone Black web server is running. You have seen that you can connect via

Secure Shell (SSH)—and that SSH server is running. Those are just a few of the processes running. I used the Linux command line to list all the processes running and found a total of 107 on my particular board while the script was running. That is a lot of stuff executing, and the script has to share with everything else that is executing. Several layers are involved in executing BoneScript, not just a simple interpreted language.

Now, one thing to consider that may speed up execution is to execute the code directly from the command line and forgo a lot of the overhead involved with using BoneScript. Maybe we can get a little more efficiency out of our program with Python.

Python—A Step Above Interpreted Language

Python is a different programming language, but still an interpreted language. We can't just drop our BoneScript code into a Python interpreter and expect it to execute. Different programming languages are written with differing syntax and ways of doing things.

For this next part, we are going to go back to the command line to install some useful tools to use Python on the BeagleBone Black provided by Adafruit (http://adafruit.com). Go ahead and log in to your board via SSH. Once logged in, we need to do a little work to add the ability to use Python. Because we use Python throughout the remainder of this book, it is a good idea to get these actions done now.

First, make sure the clock is up to date. You can do this using a utility called Network Time Protocol (NTP). From the terminal, execute the following:

```
root@beaglebone:~# ntpdate pool.ntp.org
```

This command tells the operating system to contact the server pool.ntp.org, get the time, and update the board's time. This command will likely take a moment or two. When it returns, you will see something like this:

```
10 Aug 18:40:04 ntpdate[1990]: adjust time server 50.7.0.147 offset -0.096855 sec
```

This response tells us the time the program executed, the IP address of the specific time server, and how much the clock had to be adjusted. In this case, it is about 96ms but the change can be much larger, on the order of years. Why update the clock? We are going to synchronize some libraries from the Web and we want to make sure our time is right.

The next thing you need to do is to ensure that all software is up to date. To do this, you'll use the command common among Debian Linux distributions called `apt-get`. This command makes updating your software and operating system easy. The first thing you need is an update of the libraries to which you have access:

```
root@beaglebone:~# apt-get update
```

Many lines will scroll by as the `apt` software retrieves updated library listings. This isn't a bad time to make sure you have the latest software from all of these repositories. This upgrade is accomplished by using the `apt-get` program with the `upgrade` keyword:

```
root@beaglebone:~# apt-get upgrade
```

In this case, upgrades have not been run in a while, so there is a long list of them. I entered **yes** to start the upgrade running.

```
Reading package lists... Done
Building dependency tree
Reading state information... Done
The following packages will be upgraded:
  acpi-support-base apache2 apache2-mpm-worker apache2-utils
  apache2.2-bin apache2.2-common apt apt-utils base-files beaglebone
  dbus dbus-x11 dpkg dpkg-dev gnupg gpgv libapt-inst1.5 libapt-pkg4.12
  libavcodec-dev libavcodec53 libavformat-dev libavformat53
  libavutil-dev libavutil51 libc-bin libc-dev-bin libc6:armel libc6
  libc6-dev libcups2 libdbus-1-3 libdbus-1-dev libdpkg-perl
  libgnutls26 libgssapi-krb5-2 libjpeg-progs libjpeg8 libjpeg8-dev
  libk5crypto3 libkrb5-3 libkrb5support0 liblcms2-2 libnspr4 libnss3
  libsmbclient libsoup-gnome2.4-1 libsoup2.4-1 libssl-dev libssl-doc
  libssl1.0.0 libswscale-dev libswscale2 libwbclient0 libxfont1
  libxml2 libxml2-dev libxml2-utils linux-libc-dev locales
  multiarch-support openssh-client openssh-server openssl tzdata
64 upgraded, 0 newly installed, 0 to remove and 0 not upgraded.
Need to get 49.1 MB of archives.
After this operation, 183 kB disk space will be freed.
Do you want to continue [Y/n]? Y
```

This process may take a couple minutes. Relax. Get a cup of coffee. This is a good set of commands to run to make sure everything is up to date. Any time I add software to my board via `apt-get`, I also run these two commands. You can also use the `apt-get` command to remove software and perform several other maintenance operations. A good resource to reference if you can't remember how to use an aspect of `apt-get` is the Ubuntu documentation at http://help.ubuntu.com/community/AptGet/Howto.

Now that you have your software up to date, you need to add the Python interpreter along with some other basics. Go ahead and execute the following command to add these dependencies:

```
root@beaglebone:~# apt-get install build-essential python-dev \
python-setuptools python-pip python-smbus -y

Reading package lists... Done
Building dependency tree
Reading state information... Done
build-essential is already the newest version.
python-setuptools is already the newest version.
python-pip is already the newest version.
The following NEW packages will be installed:
```

```
  python-dev python-smbus
0 upgraded, 2 newly installed, 0 to remove and 0 not upgraded.
Need to get 12.1 kB of archives.
After this operation, 116 kB of additional disk space will be used.
Get:1 http://ftp.us.debian.org/debian/ wheezy/main python-dev all \
2.7.3-4+deb7u1 [920 B]
Get:2 http://ftp.us.debian.org/debian/ wheezy/main python-smbus \
armhf 3.1.0-2 [11.2 kB]
Fetched 12.1 kB in 0s (83.9 kB/s)
Selecting previously unselected package python-dev.
(Reading database ... 59261 files and directories currently installed.)
Unpacking python-dev (from .../python-dev_2.7.3-4+deb7u1_all.deb) ...
Selecting previously unselected package python-smbus.
Unpacking python-smbus (from .../python-smbus_3.1.0-2_armhf.deb) ...
Setting up python-dev (2.7.3-4+deb7u1) ...
Setting up python-smbus (3.1.0-2) ...
```

In response, the command has checked and noted that some of the software was already installed and up to date. These packages are ignored. The only packages that need to be installed in this case are `python-dev` and `python-smbus`. With these two packages installed, you are ready to move on and install the Adafruit libraries for the BeagleBone Black. At the command line, you are going to execute a command called `pip`. This program works in a manner similar to `apt-get` but it specifically retrieves Python packages from a repository. In this case, it finds the library is already installed; therefore, no work has to be done!

```
root@beaglebone:~# sudo pip install Adafruit_BBIO
Requirement already satisfied (use --upgrade to upgrade):
Adafruit-BBIO in /usr/local/lib/python2.7/dist-packages
Cleaning up...
```

What the `pip` command suggests, as you can see, is to perform an upgrade of the library using the `--upgrade` modifier. It is executed as follows:

```
root@beaglebone:~# pip install --upgrade Adafruit_BBIO
```

This command goes off and does a lot of work. Sit back and relax again. This can take a minute or two. At the end, you will receive the following message:

```
Successfully installed Adafruit-BBIO
Cleaning up...
```

For the Python implementation, you need another library installed, called `twisted`. This library provides some timing options that will let you write the Python code in a way that is functionally similar to the BoneScript version, with a timer executing the function at one-second intervals:

```
root@beaglebone:~# sudo pip install twisted
```

Implementing Blinking Lights In Python

Now that you have the latest Python libraries from Adafruit for controlling the BeagleBone Black's GPIO ports, let's implement some alternate blinking code in Python, as shown in Listing 5.2. First, you will want to create a directory to store your code. Execute the following to create a small directory hierarchy:

```
root@beaglebone:~# mkdir bbb-primer
root@beaglebone:~# cd bbb-primer
root@beaglebone:~/bbb-primer# mkdir chapter5
root@beaglebone:~/bbb-primer# cd chapter5
root@beaglebone:~/bbb-primer/chapter5#
```

The `mkdir` command creates a new directory beneath the current directory, and the `cd` command changes directories. In this case, here are the steps you're following:

1. Create the directory `bbb-primer` in the current user (root) directory.
2. Move into your newly created directory.
3. Create the directory `chapter5` in the `bbb-primer` directory.
4. Move into the `chapter5` directory.

Now, you need to create the source code (see Listing 5.2). There are a number of ways to accomplish this task. My personal favorite is to use a separate code development environment on a desktop or laptop machine and copy the source files over to the BeagleBone Black via an SSH file transfer. I use a program called FileZilla, as seen in Figure 5.1, but there are a number of programs you can use. This is helpful because you may have many files to transfer onto your board as we progress. You simply need to select SSH File Transfer and you can log in with the root identity.

LISTING 5.2 rr_crossing_blinkers.py

```
:   """
2:  rr_crossing_blinkers.py - Python File to blink LEDs attached to
3:                             GPIO P9_12 and P9_15.
4:
5:  Example program for "The BeagleBone Black Primer"
6:  """
7:
8:  import Adafruit_BBIO.GPIO as bbb  # Declare a bbb variable
9:  from twisted.internet import task, reactor
10:
11: state1 = bbb.LOW  # Declare a var to represent GPIO P9_12 state
12: state2 = bbb.LOW  # Declare a var to represent GPIO P9_15 state
13:
14: bbb.setup("P9_12", bbb.OUT)  # Set GPIO P9_12 control to output
15: bbb.setup("P9_15", bbb.OUT)  # Set GPIO P9_15 control to output
```

```
16:
17: def blink():
18:         """ Function - blink
19:         Toggle the state variables between high and low.
20:         """
21:         global state1
22:         global state2
23:
24:         if state1 is bbb.LOW:                    # If P9_12 is LOW..
25:                 state1 = bbb.HIGH                 # ... set P9_12 to HIGH
26:                 state2 = bbb.LOW                  # ... set P9_15 to LOW
27:         else:                                    # Otherwise...
28:                 state1 = bbb.LOW                  # ... set P9_12 to LOW
29:                 state2 = bbb.HIGH                 # ... set P9_15 to HIGH
30:
31:         bbb.output("P9_12", state1)    # Update the GPIO P9_12 state
32:         bbb.output("P9_15", state2)    # Update the GPIO P9_15 state
33:
34: timer_call = task.LoopingCall(blink) # ...
35: timer_call.start(1)                  # Alternate blinking LEDs
36: reactor.run()                        # ...
37:
```

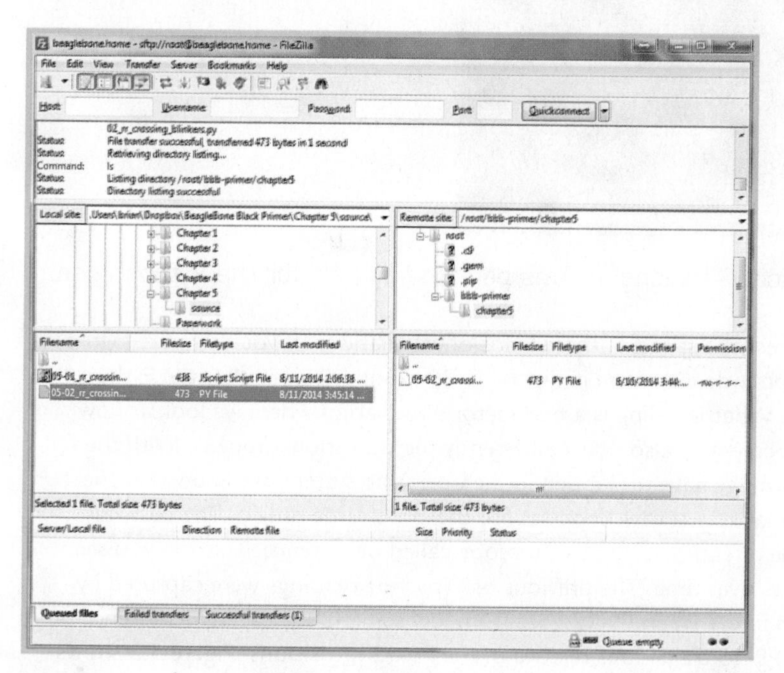

FIGURE 5.1 Using FileZilla to transfer files to the BeagleBone Black.

As you can see, this Python program is similar to our BoneScript code. In fact, all of the comments in this code, which are contained in triple-quotes (""") or after a hash mark (#) in Python, are copied from the BoneScript version. I used the external library, `twisted`, that we installed with `pip` to provide the timer functionality similar to what we had in the BoneScript version.

Once you are ready to go, you'll want to execute your Python file as follows:

```
root@beaglebone:~/bbb-primer/chapter5# python 05-02_rr_crossing_blinkers.py
```

The program will start running and you won't see any output other than the blinking LEDs. When you are ready to stop the program, press Ctrl+C. This is a universal keystroke combination that stops whatever program you have running from the command line.

With the Python script running, you can see that the lights appear to blink at exactly the same speed they did when we were running the BoneScript. However, we know this to not be the case from the last chapter. How long is the time between transitions now? Figure 5.2 shows the oscilloscope output. As you can see, the transitions are now around 0.5ms apart. Double the speed! Now, is that because Python is faster? Could be, in this instance. There is also no overhead due to the debugger or the Cloud9 IDE running. Those would put a burden on the system and cause everything to execute a little slower. However, plenty of benchmarks are available that show JavaScript (the underlying language for BoneScript) running much faster than Python under different conditions and different interpreters.

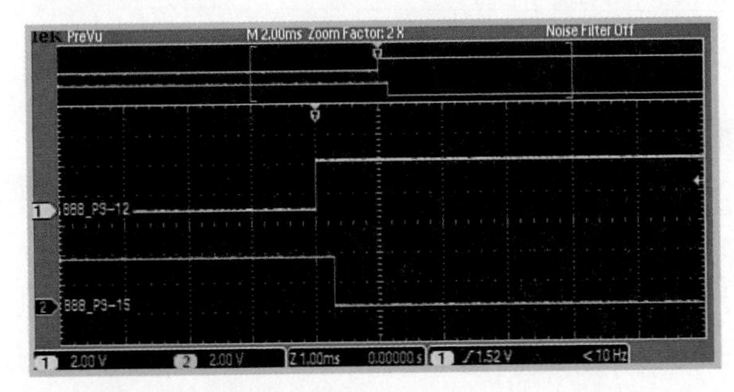

FIGURE 5.2 Oscilloscope reading at 1ms per time division for the Python code.

A lot of factors will affect the speed of these kinds of calculations. If you compare Figure 5.2 with the BoneScript code execution in Figure 4.13, it looks obvious that the Python code is running faster. What is telling is a two-factor observation where we look at how fast a transition may occur and also how consistently the transition happens. With the full knowledge that other things have to happen in the operating system, we know that the same code execution may not always happen with the exact same timing. The best way to look at this is to use a feature of the oscilloscope called persistence. This lets you see repeated measurements over time. The previous oscilloscope readings were captured by triggering the scope to freeze when P9-12 goes from low to high. By leaving persistence on, you can see all of the P9-15 transitions relative to the P9-12 transitions. Figure 5.3 shows the BoneScript version of our code executing with persistence, and Figure 5.4 shows the Python version with persistence.

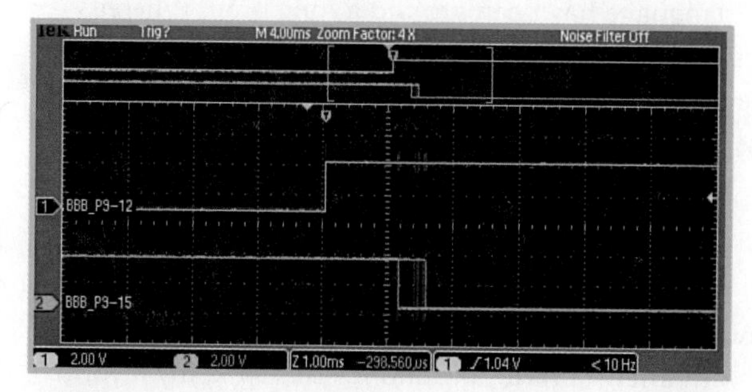

FIGURE 5.3 Oscilloscope reading at 1ms per time division for the BoneScript code with persistence.

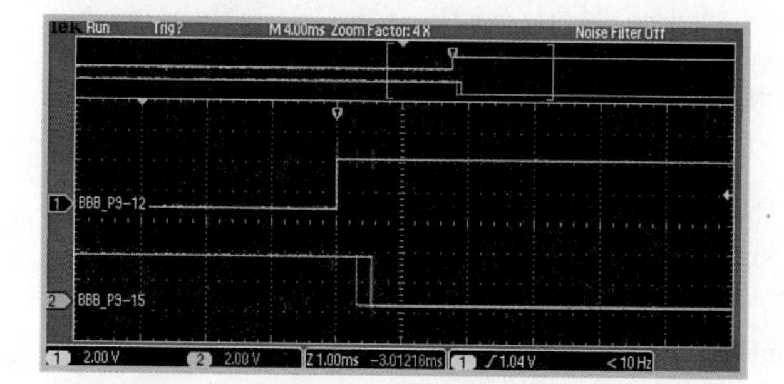

FIGURE 5.4 Oscilloscope reading at 1ms per time division for the Python code with persistence.

This is where we get into the meat of the classical difference between microcontrollers and operating system-based microprocessor systems. Having an operating system under the covers brings a number of advantages into play; however, you lose the deterministic nature of the microcontroller. If we want to get as close to the microcontroller level of speed for this type of transition, we need to look into compiled languages.

Compiled Code

A different category of source code is compiled languages. Remember that as interpreted languages are executing they are deciding, at that time, how to translate the actions into something the machine can execute. Compiled code only makes that effort once. The source code is compiled and, at that time, translated into machine language; thus, a new, executable file is created. Now, when the program executes, there is no translation. It is run directly on the hardware and scheduled into the operating system just like any other task.

Many languages can be compiled down to run natively. My personal favorite—and the language that is often used for software development and hardware interactions—is the C language or its successor, C++. The C language has been around a long time. Whereas Python has been around about 25 years and JavaScript about 20, the C programming language has been around for 45 years. Most microcontrollers and hardware-centric processors have a C or C++ compiler as part of their development environment.

As such, the BeagleBone Black has a built-in compiler as part of the Linux operating system. To go as simple as possible, let's implement our program yet again, but this time as a C program.

First, we will need a library for the C code to interact with the BeagleBone Black GPIO pins. As of this writing, I experimented with a few libraries I found for this task; my favorite (and the best performing one) I found on the Element14 website. It was developed by a user identified as shabaz, and the discussion thread where the code is discussed can be found at http://goo.gl/OgBHyO. It is a basic library but works well for this example and demonstration. The source libraries are linked on this page as zip files. To use them on your BeagleBone Black, you are first going to need to install an unzip utility with the now familiar `apt-get` function:

```
root@beaglebone:~/bbb-primer/chapter5# apt-get update && apt-get upgrade
```

You've seen these commands before. They make sure that all the information regarding the latest versions of installed software is known to the apt system and then they upgrade any packages that are out of date:

```
root@beaglebone:~/bbb-primer/chapter5# apt-get install unzip
```

This will install the unzip software. The next step is to download the library source code zip file.

```
root@beaglebone:~/bbb-primer/chapter5# wget http://goo.gl/5fYRz1


--2014-12-20 21:08:05--  http://goo.gl/5fYRz1
Resolving www.element14.com (www.element14.com)... 23.11.217.225
Connecting to www.element14.com (www.element14.com)¦23.11.217.225¦:80
... connected.
HTTP request sent, awaiting response... 200 OK
Length: 5184 (5.1K) [application/zip]
Saving to: `iofunc_v2.zip'

100%[=========================================>] 5,184        --.-K/s   in 0s

2014-12-20 21:08:05 (41.3 MB/s) - `iofunc_v2.zip.1' saved [5184/5184]
```

Now that you have the library files, unzip them into the `chapter5` directory:

```
root@beaglebone:~/bbb-primer/chapter5# unzip iofunc_v2.zip
```

```
Archive:  iofunc_v2.zip
  inflating: iofunc/iolib.c
  inflating: iofunc/iolib.h
  inflating: iofunc/libiofunc.a
  inflating: iofunc/Makefile
  inflating: iofunc/test_app.c
```

You now have an iofunc directory inside of your chapter5 directory. Go ahead and move into that directory so that we can build the library. What you've downloaded is actually source code for a C library, and you will compile it and copy the compiled binary files into the correct reference directories for linking into other programs, such as our C version of the blinker program.

```
root@beaglebone:~/bbb-primer/chapter5# cd iofunc
root@beaglebone:~/bbb-primer/chapter5/iofunc# make clean
rm -rf *.o libiofunc.a test_app core *~
root@beaglebone:~/bbb-primer/chapter5/iofunc# make all
gcc     -c -o iolib.o iolib.c
ar -rs libiofunc.a iolib.o
ar: creating libiofunc.a
gcc   -c -o test_app.o test_app.c
gcc test_app.o -L. -liofunc -o test_app
```

The two make commands invoke the information in the file called Makefile in the iofunc directory. This is a special type of file that make knows how to interpret, and it contains the information on how to compile and link the source code to create the libraries. Now that you've executed these functions, you need to copy the correct files to places in the file system where the compiler knows about to get the libraries:

```
root@beaglebone:~/bbb-primer/chapter5/iofunc# cp libiofunc.a /usr/lib
root@beaglebone:~/bbb-primer/chapter5/iofunc# cp iolib.h /usr/include
```

Now you have the library in place to interface with the BeagleBone hardware. The next step is to create the C source file for our blinking program. Go ahead and move back up a level to the chapter5 directory and create the source file in Listing 5.3.

LISTING 5.3 rr_crossing_blinkers.c

```
11:  /**
2:   * rr_crossing_blinkers.c - Python File to blink LEDs attached to
3:   *                          GPIO P9_12 and P9_15.
4:   *
5:   * Example program for "The BeagleBone Black Primer"
6:   */
7:
8:
```

```
9:  #include "iolib.h"
10:
11: #define PORT 9
12: #define PIN_A 12
13: #define PIN_B 15
14:
15: #define TRUE  1
16: #define FALSE 0
17:
18: int main() {
19:
20:     iolib_init();
21:     iolib_setdir(PORT, PIN_A, DIR_OUT);  // Set GPIO P9_12 to out
22:     iolib_setdir(PORT, PIN_B, DIR_OUT);  // Set GPIO P9_15 to out
23:
24:     pin_low(PORT, PIN_A);                 // GPIO P9_12 to low/off
25:     pin_low(PORT, PIN_B);                 // GPIO P9_15 to low/off
26:
27:     int state = FALSE;
28:
29:     while(TRUE) {
30:
31:         if (state) {         // If P9_12 is not high (is low)
32:             pin_low(PORT, PIN_A);      // ... set P9_12 to high
33:             pin_high(PORT, PIN_B);      // ... set P9_15 to low
34:
35:         } else {                      // Otherwise ...
36:             pin_high(PORT, PIN_A);      // ... set P9_12 to low
37:             pin_low(PORT, PIN_B);      // ... set P9_15 to high
38:         }
39:
40:         state = !state;
41:
42:         sleep(1);                  // Run once every second
43:
44:     }
45:
46:     iolib_free();
47:     return(0);
48: }
```

I'm not going to walk through the whole code again because I bet you're very familiar with it all by now. One thing to note, however, is the use of `#define`. This is used to define a literal text replacement and can be a good way to establish a constant without using extra memory space as the constants are substituted before the program is actually built. I use it here to define the pins and the port as a type of constant; this way, if I wanted to use other pins, I would just have to change the values here and the rest of the program will change appropriately.

Unlike the interpreted code we used before, we cannot simply execute the source file. We have to go through the previously mentioned compile process to get the executable binary. For the library we are using, we simply ran the `make` file. For our source code, we are going to compile via the command line. Make sure you are in the same directory as the C source file and then execute the following:

```
gcc 05-03_rr_crossing_blinkers.c -liofunc -o blinker
```

When this command is executed, you should see nothing. That's a good thing; after you hit Enter, the BeagleBone Black will think about it for a second and then give you back a command prompt. It is in that second that the compiler did all the work and generated a binary file. Let's look at the command, piece by piece. The last argument, `-o blinker`, tells the compiler to produce a binary executable called `blinker`. Now, you need to execute your file to see your lights blinking:

```
root@beaglebone:~/bbb-primer/chapter5# ./blinker
```

You should now see the familiar blinking of the lights. You might ask why we have to put the `./` ahead of the call to our executable. This is due to the fact that the directory we are executing in right now is not a part of our environmental PATH. I'm not going to get into all the details at this point. A Google search on the topic can give you a lot more information than I want to include here.

Now that the lights are blinking with the executing C code, we need to go back to the oscilloscope and see how the performance looks. Figure 5.5 shows the familiar 1ms scale oscilloscope view.

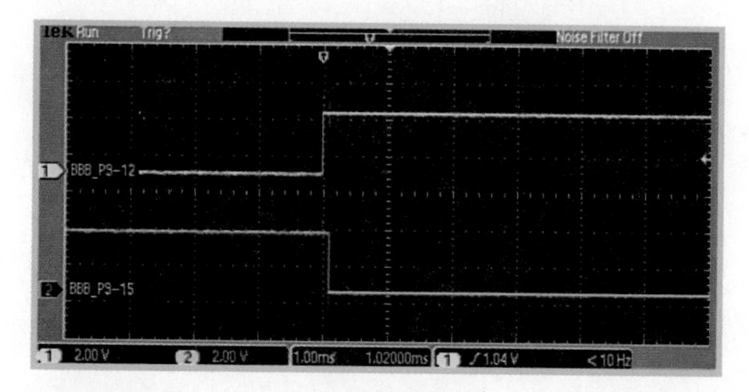

FIGURE 5.5 Oscilloscope reading at 1ms per time division for the C code with persistence.

As you can see, the transition is much more rapid and virtually imperceptible at 1ms per division. To see the difference, we need to zoom down to 4μs per division—that is, 0.000004 seconds between vertical marks. As you can see in Figure 5.6, the time delta is miniscule and doesn't waiver much at all.

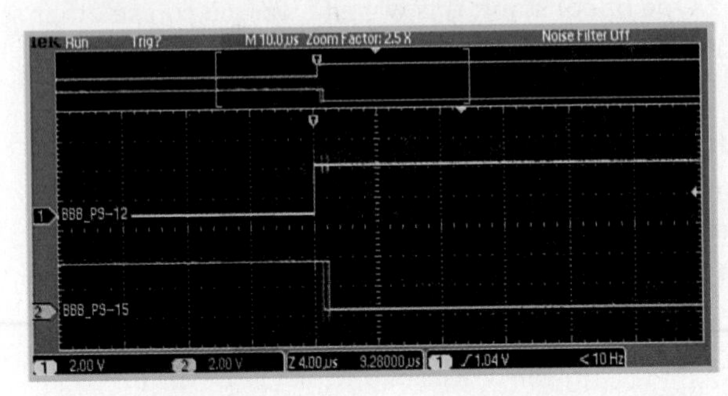

FIGURE 5.6 Oscilloscope reading zoomed to 4μs per time division for the C code with persistence.

This would be a good time to ask, "What does this all mean?" Here are a couple important points to take away:

- Compiled code will execute faster and with more precision than interpreted code. Any time you need something to happen with a lot of speed or you need to perform calculation-intensive operations, you will want to go with compiled code.
- Interpreted code is a lot faster to prototype and is more portable. I was a difficult convert from C/C++ to Python, but I can't argue with how easy it is to develop and prototype in Python. Now, BoneScript is a fine language for playing around in the system, but I write a lot more Python code than BoneScript. This is partially a personal decision. I don't like JavaScript in general. Just my thing. Nothing personal.

As we move forward, you will see a mixing of interpreted code with compiled code and an offloading of functions to hardware. This sets up the basis for our projects. We just need to look at one more fundamental: changing the BeagleBone Black's operating system.

6

Trying Other Operating Systems

So far, we have focused on the operating system (OS) that comes delivered with the BeagleBone Black. However, a number of operating systems have been ported to the BeagleBone Black. To understand what I mean by "ported," you need to reach back to our talk about the architecture of the CPU on the board, which utilizes the ARM architecture. *Ported* simply means that the operating system has been compiled to operate with a specific architecture. In this chapter, we look at the king of embedded operating systems, some of the OS options available, and the process for installing an OS onto your board.

History of the Linux World: Part I

By far, the most popular operating system for the BeagleBone Black is Linux. The default operating system as I am writing this book is a version of the Ubuntu Linux OS. You will hear about a number of different versions of Linux known as *distros,* short for *distributions*. These are different flavors of the same operating system. The core of them all is the Linux kernel. Taking a casual stroll through the lineage that brings us to the Ubuntu Linux that comes on our board today will give you a familiarity with some of the common terms you will be exposed to throughout the use of a Linux distribution.

In the beginning was ENIAC....

Okay, we won't start that far back, but the development of early computers (such as the difference engine envisioned by Charles Babbage, Colossus, ENIAC, and UNIVAC) gives us an important insight into the development of computer architectures. There were no such things as a "desktop" computer. A computer was a hulking beast of a machine, and people sought time to use the beast. As more people wanted to use *the* computer, methods had to be developed to allow for an easier interface and even for multiple people to use the system at the same time. Thus, the operating system was born.

That's the short, short version. (This is my last Mel Brooks movie reference. I promise... for now.) The early history of computer design and operating system design is fascinating and worth independent investigation. If you ever find yourself in the San Francisco Bay area, treat yourself to the Computer History Museum in Mountain View.

One of the early operating systems was Unix. Unix was initially developed at AT&T's Bell Labs, and many aspects of that first Unix system are a core part of operating systems today. The first versions of Unix were developed in the assembly language of the particular development machine, but the value of rewriting Unix in a high-level language such as C was obvious. With the operating system rewritten in C, it could be recompiled and ported to any platform that had a C compiler. This is the same idea in play when I mentioned porting operating system to the BeagleBone Black.

Unix grew, branched out, and took on a life of its own. Many operating systems today can trace their family tree to the original Unix. Do you have or have you ever used an Apple device running the OS X or iOS operating system? Those operating systems have a direct lineage to the original AT&T Unix. However, not everything that is based on Unix is derived directly from the Unix family tree.

In 1991, Linus Torvalds was working on a project to help teach himself about the Intel 80386 architecture. He based the early design on a Unix-like operating system called MINIX. He had ported the bash terminal environment and the gcc compiler and started looking for inputs, suggestions, and contributions. The 80386 processor had a lot of neat features, and it was fairly easy to write bare-bones code. In fact, all of the scientific instruments onboard the Hubble Space Telescope today are operating with bare-bones code executing on an 80386.

For a number of reasons, the world was hungry for a free and open-source Unix-like operating system, and Torvalds' small project evolved into the most ubiquitous operating system in the world. Sure, everyone knows Windows and OS X, but many people run Linux as a desktop operating system. Beyond the desktop, Linux is used in servers and in embedded systems, and is even the basis of the Android operating system.

All of the Linux flavors out there have one thing in common; the Linux kernel. The kernel is the heart of the system, and it establishes all of the core functionality of the operating system. What functionality is included after that is what differentiates the different flavors of Linux as full operating systems. Some flavors are geared toward being a desktop environment, whereas others are geared toward being embedded in a machine with the user never actually logging in to the interface.

Of course, there are other operating systems aside from Unix-based and Unix-like operating systems. The Windows platform comes to mind. The history of the modern Windows operating system traces back to an operating system called CP/M, with MS-DOS in the middle.

I would be remiss if I skipped another operating system near and dear to my heart, and to the hearts of some of my friends. The operating system is called VMS, and it is used in many legacy systems today for compatibility. It is a fun operating system that I would love to see ported to the BeagleBone Black, if for no other reason than nostalgia.

Picking an Operating System

In the world of the BeagleBone Black, we need to work with the operating systems that have been ported to the board. A great resource is the BeagleBoard.org website. If you search the projects (http://beagleboard.org/project) and filter on just the distros for the BeagleBone Black, as shown in Figure 6.1, you end up with a handy list of operating systems ported to the board.

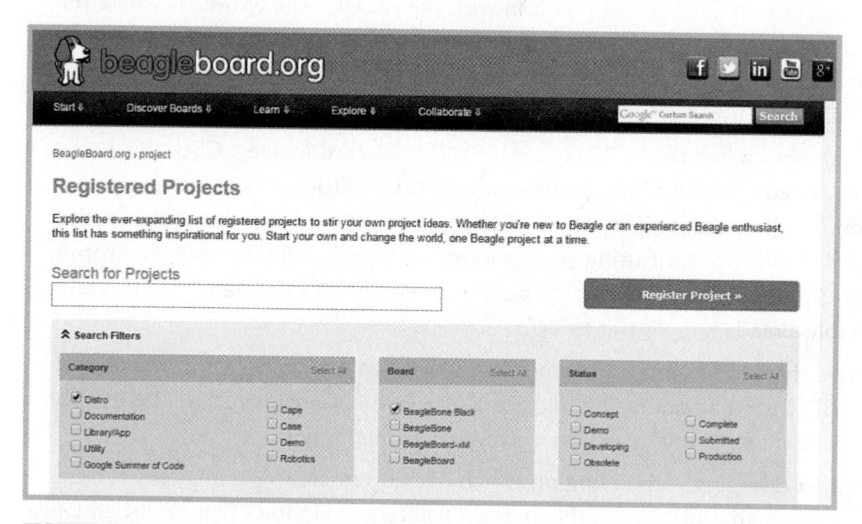

FIGURE 6.1 Searching BeagleBoard.org projects for BeagleBone Black–compatible OS distributions.

At the time of this writing, 30 distributions are listed as compatible with the BeagleBone Black. Of those 30, only four are not a Linux/Unix-type distribution. There is even one Windows Embedded port available!

A number of factors go into which distribution you would want to use. For most of the rest of this book, I will utilize the default Ubuntu distribution, but I want to show you, dear reader, how to install a different operating system. We could be boring and just load up a different, generic Linux distribution—or get even more boring and load the Windows port. Instead, let's have some fun and play some classic Super Nintendo Entertainment System (SNES) games. One of the distributions listed is the BeagleSNES project (http://www.beaglesnes.org/), a Linux version specifically designed to run an SNES emulator. Now that we have a distribution selected, I'll show you how to load it to your BeagleBone Black.

Loading the microSD Card

Now that we have an operating system selected, we need to get the image for that system. Historically, this is called a *disk image* because it is an exact, binary copy of a full file system. It is a way to take a base operating system you want to use, as installed on a system "disk," and make it available to other people. By creating a binary copy on your own "disk," you can install your "disk" in the hardware and boot the system. (I heavily used quotes around "disk" because you're almost never going to use any kind of disk-like media anymore. I honestly can't remember the last time I installed software that wasn't simply downloaded or used with a USB-type input.)

The BeagleBone Black has a handy microSD card slot that can act as a system disk. The board also has a section of flash storage built in with the eMMC. The eMMC is where the default operating system is located. We have a couple different options for where an image is stored and how we will boot it, but we can't do anything until we copy the disk image onto a microSD card.

The BeagleSNES site has a link to the "BeagleSNES microSD card image." Click this link to download the zipped image file. The download may take a little while, depending on your Internet connection, because it is over a gigabyte in size. Once you have downloaded the compressed file, uncompress it using your favorite unzipping utility on your desktop. I, personally, use 7-Zip on Windows, the utility built in to Finder on OS X—and if you're using Linux, you probably already know what to do.

Now that you have the full disk image, you need to write it to a microSD card. Make sure you have a microSD card that can fit the full size of the image. Generally, it will be 4GB or less so that it can fit in the BeagleBone Black eMMC flash memory.

Now, the roads start diverging, depending on which desktop operating system environment you are using to download and prepare the image. I will cover Windows first and then Linux and OS X together because they are very similar.

A utility is available from the Ubuntu group (creators and maintainers of the Ubuntu Linux distribution) for copying a disk image to a device, called Win32 Disk Imager. You can download this utility from a section of the Ubuntu site, **http://wiki.ubuntu.com/Win32DiskImager**. Once it is downloaded and installed, you will find the interface, shown in Figure 6.2, to be extremely straightforward.

FIGURE 6.2 The Win32 Disk Imager utility.

At this point, go ahead in insert the microSD card into whichever card reader you are using. The microSD card I bought came with a full-size SD card adapter that fits into an SD card slot on my computer. You may need an external reader/writer that connects via USB. When you have inserted the card, Windows will let you know when it has successfully opened the card and given it a drive letter. In my case, it was assigned "E:".

In the upper-left portion of the utility is a section where you can select the disk image you want to use. You can see in Figure 6.2 that I am using the BeagleSNES image I have unzipped, which is located at C:\Users\brian\Downloads\beaglesnes_full.img. Make sure you are using the .img file for whichever image you are using and not a compressed file (*.zip, *.bz, *.bz2, *.tar.gz, and so on).

Once you have set your image file, you can select the device you want to write to in the upper-right portion of the window (which is "E:" in my case). Once you have everything set, click the Copy button. The progress bar will slowly start filling as the image is written, byte for byte, to the microSD card. It could take a while. Go ahead and get yourself a snack or catch up with loved ones. When the copy is done, you will receive confirmation, as shown in Figure 6.3.

FIGURE 6.3 Win32 Disk Imager letting you know that your copy is complete.

That's all there is to it. Pop out your card, and you are ready to load it to your BeagleBone Black!

In the Linux/OS X environment, there is no need to download additional utilities. Everything is already baked in to your system. What follows is a general example on an Ubuntu Linux virtual machine. If you're a Linux/OS X user, I'm going to make the practical assumption that you can follow along and apply the instructions to your own environment. The first thing you will need to do is unzip the image:

```
brian@ubuntu:/bbb$ bunzip2 beaglesnes_full.img.bz2
```

This should return with no further information, and you should now have a file called `beaglesnes_full.img`.

Now we are going to dig into some really bare-bones commands in the *nix environment. We are going to copy the disk image with a program called `dd`. The basic description of `dd` in the *nix manual file is simple: `dd - convert and copy a file`. Simple enough, and it sounds like what we want to do. However, what it is really doing is far more powerful and can be far more dangerous. The `dd` program allows you to read and write raw data to and from input and output streams. You could, for example, accidentally write all zeros from `/dev/`

zero—a virtual file that provides a zero byte for every byte read—to your main storage device, essentially erasing the device. That could be bad, unless that is what you want to do.

In our case, we are going to copy from a file data stream to the microSD card device. To do so, we need to start with a baseline of the devices in our system so that we know which device is the microSD card when it is plugged in. To do that, we are just going to list out the space utilization for the storage devices with the `df` command:

```
brian@ubuntu:/bbb$ df
Filesystem     1K-blocks    Used Available Use% Mounted on
/dev/sda1       19478204 8222260  10243464  45% /
none                   4       0         4   0% /sys/fs/cgroup
udev              494552       4    494548   1% /dev
tmpfs             101068    1088     99980   2% /run
none                5120       0      5120   0% /run/lock
none              505340     152    505188   1% /run/shm
none              102400      44    102356   1% /run/user
```

Next, we need to plug in the microSD card we will use and then run the command again. This time, a new device should be listed:

```
brian@ubuntu:/bbb$ df
Filesystem     1K-blocks    Used Available Use% Mounted on
/dev/sda1       19478204 8222364  10243360  45% /
none                   4       0         4   0% /sys/fs/cgroup
udev              494552       4    494548   1% /dev
tmpfs             101068    1204     99864   2% /run
none                5120       0      5120   0% /run/lock
none              505340     152    505188   1% /run/shm
none              102400      48    102352   1% /run/user
/dev/sdd        15541760       8  15541752   1% /media/brian/301D-2601
```

The new item listed under the `Filesystem` column as `/dev/sdd` is our microSD card! You might find that you have something more complex, like multiple partitions on one disk, such as `/dev/sdd1` and `/dev/sdd2`. In this case, you may want to format the device with your normal formatting functionality in the operating system, just so you start on a simple footing. Likewise, you can use the `dd` command to wipe the device. A quick Google search will help you there.

Now that we know which device we have, we need to unmount it from the operating system. In Linux environments you would use `umount`; on OS X the preferred command is `diskutil unmountDisk`.

```
brian@ubuntu:/bbb$ umount /dev/sdd
brian@ubuntu:/bbb$ df
Filesystem     1K-blocks    Used Available Use% Mounted on
```

```
/dev/sda1         19478204  8222348  10243376   45%    /
none                     4        0         4    0%    /sys/fs/cgroup
udev                494552        4    494548    1%    /dev
tmpfs               101068     1204     99864    2%    /run
none                  5120        0      5120    0%    /run/lock
none                505340      152    505188    1%    /run/shm
none                102400       52    102348    1%    /run/user
```

As you can see, because we ran a `df` again right after the `umount`, the microSD card no longer shows up, but it is still physically attached as a device. Now, we can run the `dd` command to write the image file to the microSD card. Be warned, you probably have to be super-user, so you will preface with the `sudo` command. Remember, when you are using `sudo` you have a lot of privileges on the system at a very low level. With great privileges comes great responsibility.

When you execute this command, you won't see any activity in the terminal until the write is complete. It could also take a good bit of time. Relax. Get up, grab a coffee or beverage of your choice, and make contact with your loved ones.

```
brian@ubuntu:/bbb$ sudo dd bs=4M if=beaglesnes_full.img of=/dev/sdd
[sudo] password for brian:
945+1 records in
945+1 records out
3966763008 bytes (4.0 GB) copied, 883.681 s, 4.5 MB/s
```

That's all we have to do in the *nix environment.

Now that you have an image to boot, you can take the microSD card, put it into your BBB, and boot it. When you power on, the beagleSNES image is designed to make sure the system boots from the image on the microSD card and not the eMMC. You will need to connect your BeagleBone Black to a display, such as a monitor or television, using the micro-HDMI connector. Figure 6.4 shows my configuration with the default screen and the art changed for one of my favorite games of the classes SNES system, *F-Zero*. I have a micro-HDMI cable connected to a monitor, a Logitech USB Gamepad connected via USB, power, Ethernet, and the microSD card with the beagleSNES image installed.

Figure 6.5 gives a better look at the BeagleBone Black configuration and the microSD card plugged in.

Now, if you're all about the SNES and want to play for a while, it is time to start digging through the beagleSNES documentation on how to modify the configuration files and add game files. Also, I should note here that downloading the games files for games you don't actually own is illegal. Make sure you own your games!

FIGURE 6.4 beagleSNES booted from my BeagleBone Black with a USB controller connected.

FIGURE 6.5 Zoomed in view of the BeagleBone Black with the location of the microSD card indicated.

What we've achieved here is a fun example of installing a different operating system environment on your BeagleBone Black. Here is a recap of the process:

1. Use a different machine to find the BeagleBone Black–compatible operating system image file you want to use.
2. Make sure you unpack or uncompress the file if it is in a .zip format or uses other compression.
3. Write the image to a microSD card that has the space for the image file. Keep in mind, the image files can be very large. The beagleSNES image file is 4GB. Remember, on Windows you will use the Win32 Disk Imager and on a *nix system, you will use `dd`.
4. Insert the microSD into the slot on the BeagleBone Black.
5. Power up!

These are the most general directions possible, and not every system is built the same way. For example, the beagleSNES boots very happily straight from the microSD card. Some other images may not have been built the same way and will require you to press the Boot switch on the board when the power is applied in order for the board to know to boot from the microSD. This is another place where it is crucial that you read the documentation for what you are doing. The beagleSNES documentation will walk you through all of the steps in detail if you run into a problem.

Let me repeat myself here: Make sure you read any directions and documentation. The nature of the BeagleBone Black landscape is *change*. New versions of the various operating systems are released on a regular basis, and completely new distributions are being created by individuals and teams all the time. If you followed all of the preceding directions to install the latest Debian release, the board would actually boot and show activity while it overwrites the eMMC image—a process that can take tens of minutes. Other distributions require you to press the Boot Switch button near the microSD card slot. Read the documentation, and you should have your pain reduced.

Because the beagleSNES image doesn't overwrite the eMMC, at least as of this writing, you can power down the board, remove the microSD card, and power on again to bring up the normal Debian environment. Actually, if you leave your monitor or display plugged in, you will see a normal desktop-looking environment launch.

This represents the end of what we can call the "basic" training for the BeagleBone Black board. In the next chapter, we start to delve into what we can do with the board and some key skills for building larger projects.

Expanding the Hardware Horizon

We've developed an excellent set of the fundamentals of a BeagleBone Black so far. At this point, you know how to log in, move around in the Linux environment, write basic software, and upgrade the operating system. Now it is time for us to delve deeper into how to use some of these capabilities. It is time to communicate with the expansion headers by doing more than just blinking a pair of LEDs.

In Chapter 5, "A Little Deeper into Development," we wrote a basic program to blink back and forth on a pair of lights, much like a railroad crossing. This signal is a good warning, but doesn't really give any kind of meaningful information. What if we wanted to present a number to the user via the outputs? The binary representation of numbers relies on the state of each binary digit—also known as a *bit*—being 0 or 1. This can also be seen as on or off, true or false, and high or low. This makes it very easy to represent a number with LEDs! In this chapter, you learn how to expand the hardware to add new functionality to the BeagleBone Black.

Binary Basics

You've probably been able to count for a very long time, but if you've never counted in binary, it can be a little confusing at first. Let's start counting and work toward a better understanding of binary numbers. A good place to start is at zero.

A zero is a zero is a zero. Adding one is easy because we know what one looks like in decimal. In binary, we already established that a bit can be either 0 or 1. So counting from 0 to 1 in binary is exactly the same in decimal. What happens if we add one more? Decimal goes to 2, but there is no 2 in binary! We are still counting with a number set, just a limited number set. What happens in decimal counting if we get to 9, the highest digit we can represent? We add another digit place and add a 1. We add a 1 to the "tens" place every time we hit a 9 in the "ones" place until we hit 99. What do we do then? Add a digit at the "hundreds" place and keep going! What are we really doing here? A quick reminder of basic math at what underlies the decimal numbers we deal with every day. If we break a decimal number down, let's say 255, it is actually a sum, 200 + 50 + 5. Each digit represents a multiple of the value of 1 in that position. Table 7.1 illustrates this.

TABLE 7.1 Decimal Counting Basics

2			5			5		
2	*	100	5	*	10	5	*	1
200		+	50		+	5		
255								

And what are 100, 10, and 1? They are powers of the numeric basic of decimal counting, 10! (see Table 7.2)

TABLE 7.2 Decimal Counting Made Overly Complicated

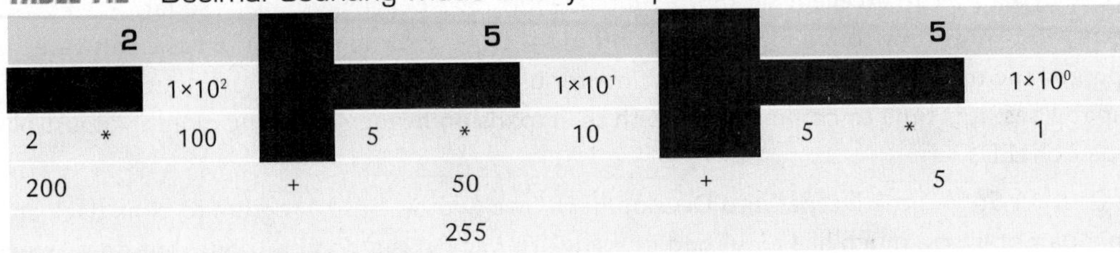

2			5			5		
		1×10^2			1×10^1			1×10^0
2	*	100	5	*	10	5	*	1
200		+	50		+	5		
			255					

Of course you don't think about numbers like this every day. Decimal counting is a normal, everyday task that you can accomplish in your head without worrying about the specifics of the multiples of 10 of each digit. Binary counting, however, is something to which people often aren't accustomed. This little trip down numeric foundations lane gives us a toolkit to understand binary a little better, as it is going to become an important foundation. Let's look at a three-bit number in a similar fashion to the way we looked at decimal 255. Remember, your numeric basis for binary is 2, not 10! We will look at the binary number 101 and find the decimal representation (see Table 7.3).

TABLE 7.3 Binary Counting Made Overly Complicated

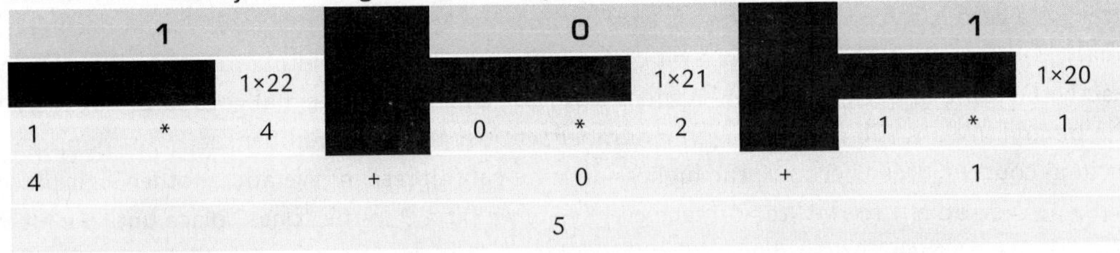

1			0			1		
		1×22			1×21			1×20
1	*	4	0	*	2	1	*	1
4		+	0		+	1		
			5					

Simple! In computing, numbers are collected into groups of eight bits, known as a *byte*. What is the highest number representable with eight bits? The highest number we could represent with eight digits in decimal is 99999999, the maximum value in each digit. So, the highest number you can represent with eight-bit positions is 11111111. What is this number equivalent in our nice, comfortable decimal form? Check out Table 7.4 to work it out.

TABLE 7.4 Binary Counting Made A Little Easier

Bits (Binary)	1	1	1	1	1	1	1	1
Bit Max (Decimal)	128	64	32	16	8	4	2	1
Bit × Max (Decimal)	128	64	32	16	8	4	2	1
Sum (Decimal)	255							

So, the highest number you can represent with eight bits is decimal 255! Just to make sure we are clear, let's try another binary number where we randomly choose the value of each bit position and figure out the decimal equivalent of that number (see Table 7.5).

TABLE 7.5 Binary Counting Made A Little Easier

Bits (Binary)	0	1	1	0	0	1	1	0
Bit Max (Decimal)	128	64	32	16	8	4	2	1
Bit x Max (Decimal)	0	64	32	0	0	4	2	0
Sum (Decimal)	102							

So, the binary number 01100110 is 102 in decimal. Now, we are ready to represent some numbers with our LEDs.

Hardware Representation

First, you need to break out control for the LEDs from the GPIO pins. You will need eight pins for control. You already know about P9_12 and P9_15 as good pins. You can actually use P9_12 through P9_17 for the first six control lines. Skip P9_17 through P9_20 because although they are GPIO pins, they are actually reserved for other uses. You will get a little better understanding of this in a few paragraphs. With the skip you can then use P9_21 and P9_22 to round out the eight bits. Toggle these between high and low to represent 1 and 0, respectively. So how does this circuit look? Take a look at Figure 7.1. For this circuit, you will need the following:

- Eight LEDs
- Nine jumper wires to go from the BeagleBone Black to the breadboard.
- Eight resistors with a resistance of at least 330 Ω to keep the current draw in check. (I used 680 Ω to be cautious. Probably overly cautious, but it won't hurt.)
- Eight more jumpers to bring the individual LEDs back to ground.

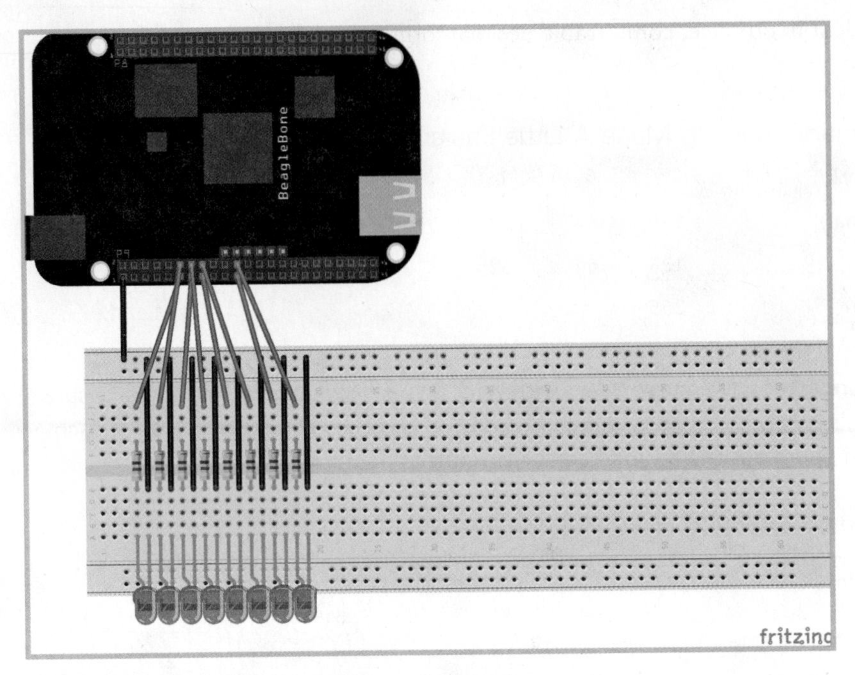

FIGURE 7.1 A circuit to illuminate our bits.

Now, you need some source code that can drive this circuit. What do you want the program to do with the LEDs you have? You will count from 0 to 255 in binary. This should be simple enough, right? Listing 7.1 is our counting program in Python.

LISTING 7.1 binary_counter.py

```
01: """
02: binary_counter.py - Illuminate 8 LEDs as binary number representation.
03:
04: Example program for "The BeagleBone Black Primer"
05: """
06:
07: import Adafruit_BBIO.GPIO as GPIO
08: import time
10:
11: PINS = ['P9_22', 'P9_21', 'P9_16', 'P9_15',
12:         'P9_14', 'P9_13', 'P9_12', 'P9_11']
16:
17: def check_bit(number, bit):
18:     return (number & (1 << bit)) != 0
19:
```

```
21: for pin in PINS:            # Configure all the pins to output & off
22:     GPIO.setup(pin, GPIO.OUT)
23:     GPIO.output(pin, GPIO.LOW)
24:
25: print 'Binary Counter - <ctrl>-c to exit.'
26:
27: try:
28:     for i in range(MAX_NUMBERS):
29:         for bit, pin in enumerate(PINS):
30:             if check_bit(i, bit):
31:                 GPIO.output(pin, GPIO.HIGH)
32:             else:
33:                 GPIO.output(pin, GPIO.LOW)
34:
35:         time.sleep(DELAY)
36:     time.sleep(5)
37:
38: except KeyboardInterrupt:
39:     pass
40:
41: finally:
42:     print 'Done! Clearing bits...'
43:     for pin in PINS:
44:         GPIO.output(pin, GPIO.LOW)
45:     GPIO.cleanup()
46:     sys.exit(0)
```

This code is a little more complex than you've seen before, so we will look at it a little differently. The first line should be familiar to you now; it describes the program and imports any libraries we will need.

Lines 10 and 11 define a list of pins as strings, with the rightmost LED's pin being the first (or zeroth) item in the list:

```
10: PINS = ['P9_22', 'P9_21', 'P9_16', 'P9_15',
11:         'P9_14', 'P9_13', 'P9_12', 'P9_11']
```

Why would we do this? Because we want to have the least-significant bit to the right, and the first item in the list represents our least-significant bit. The *least-significant bit*, or LSB, is the bit that, when changed, changes the overall value of the number with the least amount. In normal numbers, the way we traditionally write them out, this is the digit or bit furthest to the right. The *most-significant bit*, or MSB, is the bit that changes the value the most, which is the digit or bit furthest to the left traditionally.

For example, say you have the decimal number 28. If you change the 8 by adding 1, you've only changed the overall value of the number by 1, the least significant change for that number. If you change the 2 by adding 1 to that digit, you've changed the value by 10, the most significant change for that number. The same goes for binary numbers. Say you have the binary number 110 (decimal 6). If you change the 0 in the bit furthest to the right, you've only changed the value by 1 to binary 111 (decimal 7), but if you change the 1 furthest to the left you've changed the value to 010 (decimal 2), a change of 4. This becomes important later because sometimes the LSB is represented to the right, and sometimes to the left, depending on the direction of data flow and the protocols in place. In the circuit laid out in Figure 7.1, the LSB is the LED furthest to the right:

```
13: def check_bit(number, bit):
14:     return (number & (1 << bit)) != 0
```

Lines 13 and 14 bring another important concept in programming, and the implementation in Python, to light—the idea of program functions. A function is useful to perform some kind of action that needs to be executed repeatedly without having to have another copy of the code at every place. The `def` statement says that you are going to *define* a function and call it `check_bit`. The items in the parentheses tell you that the function takes two arguments, called `number` and `bit`. The *scope* of these two arguments is only within the execution of the function. Think of it this way: When the function is called, you can, with few exceptions, forget about the rest of the program, because the function only knows the variables it has been fed as arguments. The `return` statement tells you that you are going to pass the value of whatever comes after the statement back to the place where the function was called.

This is a lot like algebra and the infamous y = f(x). In that case, f is the name of the function and x is the argument. When the function f is evaluated, the result is passed to y. Because you have two variables, in this case, it is more like a three-dimensional function, z = f(x, y). So what is all the stuff after the `return` that is being evaluated? This is the introduction of bitwise logic.

So, let's say the values passed into this function are decimal 2 and 0. In binary, decimal 2 is 10, and you should think of the `number`, the variable, in binary terms here. Just like in algebra, whatever is in the parentheses is evaluated first. In this case, in the innermost parentheses is the statement `(1 << bit)`. In this case, the << is an operator, just like plus, minus, divide, and multiply. It is a special operator known as a *bitwise shift left*. A bitwise shift left takes a number's binary representation and shifts all the values left by however many places are stated on the right side of the operator. Now, remember, a 1 is a 1 is a 1. So a binary 1 and a decimal 1 are the same thing. In an eight-bit environment, such as our LED display, a 1 can be represented as 00000001 with no shift. Therefore, the resulting number is 00000001. What if you passed a value of 1 to `bit` when you called the function? The result of this expression would be 00000010. See, you moved the 1 to the left by one! The operator fills in a 0 automatically in the rightmost position. If you had the value of decimal 128, or 10000000, and shifted left by one, you would lose the value of the number completely as the MSB is shifted off to the left into the vast bit-bucket in the sky and a 0 fills in the LSB position.

So for the basic example of the binary `number` 10 (or 00000010) and `bit` 0, you can now replace the inner parentheses with the binary value of 00000001. So, what about that ampersand in the outer parentheses? Another bitwise operator known as *and*. This operator produces a new number based on the combined value of the number on either side of the operator, bit by bit. It returns a 1 in each bit position where the bit positions at the same spot in *both* numbers is a 1, and returns 0 otherwise. Check Table 7.6 for something called a *truth-table* for a one-bit example.

TABLE 7.6 Truth Table for Bitwise AND (&)

Left Side		Right Side	Result
0	&	0	0
1	&	0	0
0	&	1	0
1	&	1	1

Pretty simple. If the left *and* the right aren't both 1, then the value is 0. So what about a pair of bigger numbers? Say, 1101 and 0101? You go bit by bit through the numbers and check with a result of 0101, as shown in Table 7.7.

TABLE 7.7 Bitwise AND (&) for Bigger Numbers

First Number	1	1	0	1
Second Number	0	1	0	1
Result	0	1	0	1

So, back to line 14, you are now at binary 10, or 00000010 for `number` and binary 00000001 for the substituted value of the inner parentheses. What is the result of a bitwise AND of these two numbers? It is 00000000. Zero! Zip! Zilch! So the final part of the value is a Boolean check. This kind of operator returns a logical `True` or `False`, two useable keywords in Python, but not all languages. The `!=` operator means "not equal." Try to stick with me here to avoid the double negative. If the two values are *not equal*, the result is `True`; otherwise, it is `False`. What you are doing here is always checking to see if the value of the bitwise operations result in 0. In this case, the result of the bitwise operations is 0, and you are comparing against 0, because the two are, indeed, equal. Therefore, you return a `False`! Table 7.8 walks you through the whole sequence with our sample values.

TABLE 7.8 Stepping Through `check_bit`'s Evaluation

(	number	&	(	1	<<	bit	)	)	!=	0
(	number	&	(	1	<<	0	)	)	!=	0
(	number	&	(	00000001			)	)	!=	0
(	number	&	00000001					)	!=	0
(	00000010	&	00000001					)	!=	0
(	00000000							)	!=	0
00000000									!=	0
False										

Just to make sure it is clear, Table 7.9 does the same thing but with `bit` set to 1.

TABLE 7.9 Stepping Through `check_bit`'s Evaluation with `bit` Set to 1

(	number	&	(	1	<<	bit	)	)	!=	0
(	number	&	(	1	<<	1	)	)	!=	0
(	number	&	(	00000010			)	)	!=	0
(	number	&	00000010					)	!=	0
(	00000010	&	00000010					)	!=	0
(	00000010							)	!=	0
00000010									!=	0
True										

So what is this function doing? It is taking a number and a bit within the number (LSB first, at position 0) and checking to see if the value of the bit is 1 or 0. Note that in Python, as in most programming languages, you don't *need* to do the Boolean `!=` check. A value of 0 will always be considered `False` and any other value is `True`. Because a value of anything other than 0 can occur only when the bit isn't set, you can just return the value of the bitwise operations. However, I think it is important for source code to be clear and want to emphasize that we don't care about the actual *value* of the bitwise operation, just a `True` or `False` if the particular bit is set.

From here on out, I am going to make it easy to determine whether I am talking about a decimal representation or binary representation by preceding binary numbers with the characters 0b. So if you see 100, you know I mean decimal 100. If you see 0b100, you know I mean the binary version with the decimal value of 4.

Another important concept presented here is an understanding of why the first item in the list on lines 10 and 11 of Listing 7.1 has an index of zero. List or array indexes start with 0 and the last element will always be the length – 1. This is because you aren't selecting an element, you're defining an offset. Let's say I had the following line of code defined:

```
my_list = [a, b, c, d]
```

The variable `my_list` actually points to the beginning of the list; if you are defining an offset from the start, you don't want to move anywhere at all! This leaves you with an offset of 0. Now, `my_list` has a length of 4 but what is the offset of the last element? It is the start + 3. Hopefully, following Table 7.10 will clear this up for you.

TABLE 7.10 Array/List indexing example

my_list[0]	my_list[1]	my_list[2]	my_list[3]
A	B	c	d

```
16: for pin in PINS:                    # Configure all the pins to output & off
17:       GPIO.setup(pin, GPIO.OUT)
18:       GPIO.output(pin, GPIO.LOW)
```

Lines 16 through 18 create a loop through all the elements of `PINS` to configure everything for a startup. Line 16 says the following, in English: "The next couple lines are going to be a loop, and I want you to take the values in the list `PINS` and every time the loop executes assign the next item, in order, to the `pin` variable and only execute for the number of items in the list `PINS`." As it goes through each pin, it sets the pin to output and sets the pin to off, or logic level low. Easy! Line 20 prints a message to the User's console to let the person running the program know what is going on:

```
22: try:
...
33: except KeyboardInterrupt:
...
36: finally:
```

Now, in a clump, you will look at lines 22, 33, and 36: a `try` block. The `try` statement on line 22 opens the block to define the code you want to attempt to execute. The statement on line 33 says that you need to stop executing the code in the `try` section if a `KeyboardInterrupt` (Ctrl+C) is encountered. This is a nice clean way of stopping the program early. The finally statement on line 36 executes if the `try` portion completes *or* the `except` section is invoked, which is code that is executed no matter what. This bit of code in `finally` simply prints a message letting the user know the program is done, turns off all of the pins, and runs the GPIO library's cleanup functionality. This is all an improvement over our original blinker program because it allows for a clean stop to the program.

```
23:       for i in range(256):
24:             for bit, pin in enumerate(PINS):
```

```
25:                     if check_bit(i, bit):
26:                             GPIO.output(pin, GPIO.HIGH)
27:                     else:
28:                             GPIO.output(pin, GPIO.LOW)
29:
30:             time.sleep(0.1)
31:     time.sleep(5)
```

Lines 23 through 31 are the real meat of the program. Stepping through the program, it sets up a loop that executes 256 times (0 through 255). Each time through the loop it starts another loop, this one to run through each pin in `PINS` and also assign a bit number as a function of the index of the list. Next, we call the `check_bit` function that we have gone through very thoroughly and check each bit in the number from the top loop, stored in the variable `i`. If the bit is set, the corresponding LED is turned on; otherwise, the LED is turned off and then waits a tenth of a second before going to the next number—just enough time to see the changes but short enough that it doesn't take a long time to execute. Finally, after all the looping is done and we have all the LEDs turned on at 0b11111111, the program pauses for five seconds to let all eight LEDs glow in glory. Figure 7.2 shows a photograph of the counter hardware executing the program, stopped at 0b00000010, or 2.

FIGURE 7.2 Breadboard implementation of our binary counter.

While on the surface this appears to just be an introduction to binary counting and bit operations, there's actually more than meets the eye. It is an introduction to a communications style called parallel communications. Remember those old ribbon cables?

If you're too young to remember ribbon cables, go to a computer museum and you'll find an example there. Those ribbon cables are a wire harness to carry multiple data lines together for a parallel data path.

Another way to think about this program is that you could have another piece of hardware attached to the eight pins and provide that device with a number from your BeagleBone Black. Figure 7.3 shows a capture from a logic analyzer, decoding the eight bits as a number every 0.1 seconds.

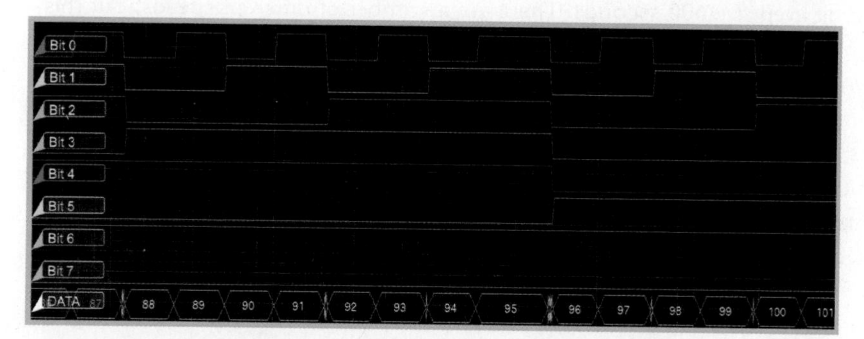

FIGURE 7.3 Logic analyzer view of our program executing, sending a sequence of numbers across eight data lines.

The use of single pins for attaching to a sensor, a single indicator light, and other options is illustrated in later chapters. A number of your interactions will be with other hardware with some kind of smarts built in. If you had to make a parallel connection to multiple devices, you would use up pins on the BeagleBone Black pretty quickly and things would get pretty messy, as you can see in Figure 7.2. There is a reason you don't see big ribbon cables inside computers anymore. Messy isn't bad when you are prototyping, but overly complex solutions aren't usually the answer either. Parallel has its place, but it is usually within a single electronics board where high-speed communications and GHz clock speeds are the norm. A simpler way that's only a little slower than parallel but far more effective and efficient from a complexity standard is *serial communications*.

Serial Communications

Serial communications is something that predates computers and has its origins in telegraphy. Either the line is sending a signal or not. Sound familiar? Like a bit, it's on or off. Follow that through many generations of improvements and you have the basis for modern, serial communications. In this chapter, we will look at the granddaddy of serial protocols: Universal Asynchronous Receiver/Transmitter (UART).

UART, pronounced "You-Art," is probably the simplest serial interface to understand. It works on one basic principle: both the transmitter and the receiver agree on how fast data will be transmitted across the connection and how much data will be transmitted at a time.

One wire is used to transmit and another wire to receive messages back if the connection is two-way. The signal on the wire sits high most of the time until a signal start is indicated, and then the transmitter transitions the signal low to indicate that it is ready to transmit. When the receiver sees the signal go low, it starts looking for the agreed-upon amount of data at the rate that is expected. When the transmitter is done, it takes the line high again. That's a lot of information, so we will step through it with an example.

First, the speed. For this connection, the agreed-upon speed, known as *baud*, is 9600 bits per second, or a bit every 1/9600 seconds. There are a number of different speeds, but this one is pretty common for short serial connections like this. Next, there is the concept in UART of the *start bit*. When the signal initially goes low, there is a bit where the line holds low to make sure the receiver is ready. The start bit is always present in a transmission. Next, both sides need to agree on how many bits of data are in the transfer. In most cases, this is 8 bits, but a couple other options are available. Also, this is generally defined so that the LSB is the first bit transmitted. Sometimes something called a *parity bit* can be added after the data section. Parity is a type of error check and can be either even or odd. More on that in a moment. Finally, just like the start bits, you can have stop bits defined. Start bits are logic low, and stop bits are logic high.

For the example in Table 7.11, the baud rate is set to 9600, the parity bit is present and set to even, and there is one stop bit. The data payload is the byte 0b010000001. Remember, the LSB will go out in the stream first!

TABLE 7.11 Sample UART Data Stream for the Byte 0b01000001 at 9600 bps, a Start Bit, Even Parity, and One Stop Bit

Description	Start Bit	Data Bit 0	Data Bit 1	Data Bit 2	Data Bit 3	Data Bit 4	Data Bit 5	Data Bit 6	Data Bit 7	Parity (Even)	Stop Bit(s)
Logic Level	0	1	0	0	0	0	1	0	1	0	1
Time From Start (ms)	0	0.10	0.21	0.31	0.41	0.52	0.63	0.73	0.83	0.93	1.04

Based on everything we've done in this chapter, you'll likely feel good about everything except the parity bit. The parity bit is a quick way to check for an error in the data. In our case, we set the parity to even. This means that there should be an even number of bits between the start bit and the stop bit, *including* the parity bit. In the case of our examples, two bits are set to 1, so there is no need for the parity bit to be set. What if we had set the parity to odd? Then the parity bit would have been set to 1 so that there are three bits (an odd number of bits) in the data package. This is useful as a quick check for whether a bit may be out of place. If the number of bits received is odd and you are expecting even, you know you encountered a problem.

Inspecting UART

Let's look at this with a real-world example. Figure 7.4 shows a capture from a logic analyzer of a UART transmission of the same number we used in Table 7.8, 0b01000001. The communications for the device the BeagleBone Black is connected to is 9600 bps, no parity bit, one stop bit.

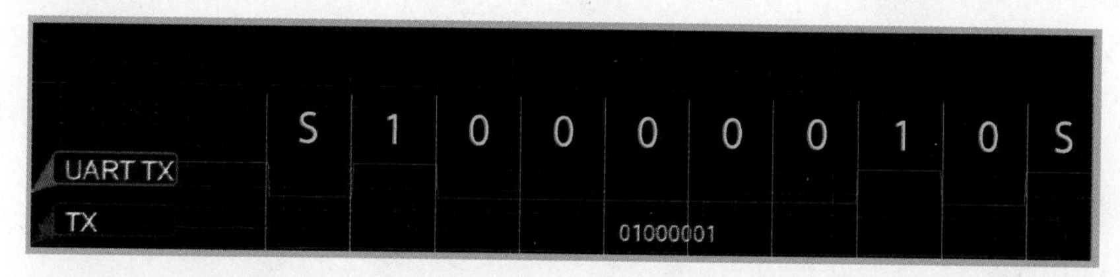

FIGURE 7.4 View of a UART serial connection with a transmission of 0b01000001.

Note that the logic analyzer provides a translation of the data payload only on the blue TX line. UART TX is the raw logic capture. The translation is shown as we would normally read it, with the LSB to the right; however, in the time domain of the analyzer, the sequence is shown with increasing time. The LSB is sent first, so the LSB is to the left on the UART TX line. Note the start bit at the front and how the line is held high at the beginning and end.

The device I am communicating with is a SerialLCD display from SparkFun that is compatible with the BeagleBone Black's 3.3V logic levels. It is devices like this that I expect you will see most often for projects with bits and pieces using the raw GPIO power. This is the power of a board like the BeagleBone Black—its ability to communicate easily with other devices and to allow for quick prototyping. The SparkFun SerialLCD is particularly neat because you can send it text, like the number in the preceding examples, and it will be displayed. Yes, the number can be a character of text. A few standards are around where everyone has agreed on what characters the numbers represent. Remember, everything is a number to a computer!

So, what is the decimal equivalent of 0b01000001? It is 65, and if you look 65 up in an ASCII table, it is the letter *A*. And sure enough, the letter *A* appears on the display, as shown in Figure 7.5.

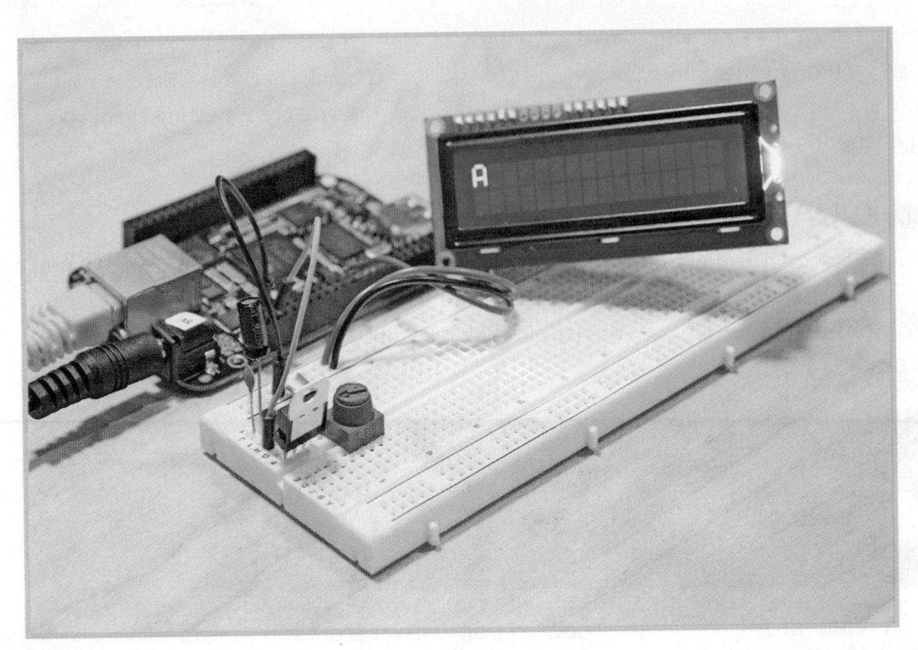

FIGURE 7.5 Connected serial display with a lot less pins from the BeagleBone Black.

Notice all those extra components to the left of the breadboard? Those components provide a 3.3V power supply to the SerialLCD board. The 3.3V outputs on the BeagleBone Black (P9_3 and P9_4) do not have the capability to provide enough power to run the board and the backlight. However, the 5V supplies on the P9_5 and P9_6 pins are driven directly by the input power jack. Plenty of power is there as long as that power source to drive the board is used. A component called a *voltage regulator* (in this case a variable voltage regulator) is used. Figure 7.6 shows the schematic for the section of the breadboard containing the regulation and a layout for the components. To better understand this circuit, look up the voltage regulator's datasheet. The specific regulator used is a LM317 variable voltage regulator. The datasheet is always important to read when you are using a device and will help prevent damage to components and devices.

In order to send the message, the Adafruit_BBIO library is used to enable serial communications. This is where a discussion of the multiple uses of a pin comes into play. The built-in UARTs are connected to specific pins with specific functionality on those pins. The Adafruit_BBIO library allows you to select one of five built-in UARTs. Table 7.12 is an update of Table 2.2 from Chapter 2, "Introduction to the Hardware," where we laid out the high-level pin configurations. This version updates to show the UART pins, highlighted in yellow. You'll notice that multiple pins are associated with each of the five UARTs. RX is the abbreviation used for receiver, TX is the abbreviation for transmitter, CTS is the Clear To Send signal (used in the synchronous version of the protocol), and RTS is Ready To Send and is also used for synchronous communications.

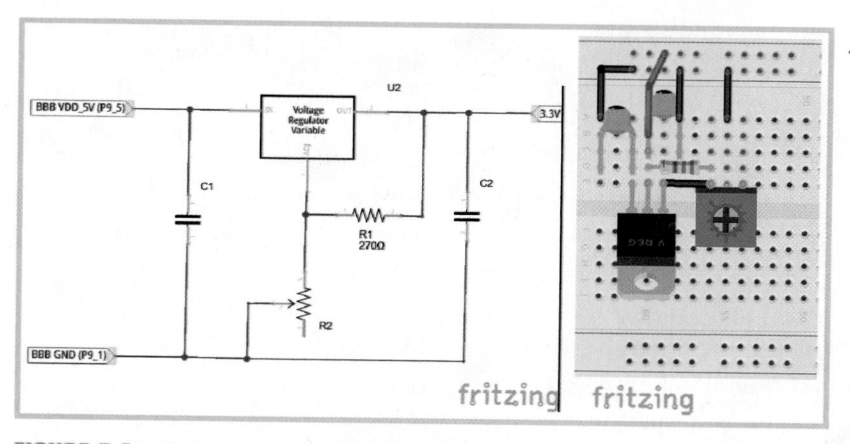

FIGURE 7.6 Schematic and layout for a 3.3V voltage regulation circuit to supply 3.3V power from the 5V supply pins.

TABLE 7.12 BeagleBone Black Expansion Headers

P9				P8			
Ground	1	2	Ground	Ground	1	2	Ground
+3.3V Power	3	4	+3.3V Power	GPIO	3	4	GPIO
+5V Input Power	5	6	+5V Input Power	GPIO	5	6	GPIO
+5V System Power	7	8	+5V System Power	GPIO	7	8	GPIO
+5V Logic Level	9	10	System Reset	GPIO	9	10	GPIO
UART4 RX	11	12	GPIO	GPIO	11	12	GPIO
UART4 TX	13	14	GPIO	GPIO	13	14	GPIO
GPIO	15	16	GPIO	GPIO	15	16	GPIO
GPIO	17	18	GPIO	GPIO	17	18	GPIO
UART1 RTS	19	20	UART1 CTS	GPIO	19	20	GPIO
UART2 TX	21	22	UART2 RX	GPIO	21	22	GPIO
GPIO	23	24	UART1 TX	GPIO	23	24	GPIO
GPIO	25	26	UART1 RX	GPIO	25	26	GPIO
GPIO	27	28	GPIO	GPIO	27	28	GPIO
GPIO	29	30	GPIO	GPIO	29	30	GPIO

P9				P8			
GPIO	31	32	ADC Ref Voltage	UART5 CTS	31	32	UART5 RTS
Analog Input	33	34	Analog Ground	UART4 RTS	33	34	UART3 RTS
Analog Input	35	36	Analog Input	UART4 CTS	35	36	UART3 CTS
Analog Input	37	38	Analog Input	UART5 TX	37	38	UART5 RX
Analog Input	39	40	Analog Input	GPIO	39	40	GPIO
GPIO	41	42	UART3 TX	GPIO	41	42	GPIO
Ground	43	44	Ground	GPIO	43	44	GPIO
Ground	45	46	Ground	GPIO	45	46	GPIO

As you can see, when you use certain UART options, pins become unavailable. If you tried to put a SerialLCD connection on UART2 or UART4 and have the binary counter program connected and operating at the same time, you would run into problems. This is why pin selection is important. For completeness, Listing 7.2 shows the source code used to run the SerialLCD and send characters. The code makes use of a Python serial library to handle the interaction with the UART. Comments are intentionally left out. However, you can look up the datasheets for the SparkFun SerialLCD as well as check the documentation for the pySerial library, the source of the serial data libraries, and look up the Adafruit BBIO libraries to better understand the way the libraries interact.

LISTING 7.2 serial_output.py

```
01: import Adafruit_BBIO.UART as UART
02: import serial
03:
04: UART.setup("UART1")
05: tty1 = serial.Serial(port="/dev/tty01", baudrate=9600)
06:
07: tty1.open()
08: tty1.write('FE01'.decode('hex'))
09: tty1.write('A')
14: tty1.close()
```

In the next chapter, we are going to get serious about expansion and fly around with Capes!

Low-Level Hardware and Capes

We've spent a good portion of our journey getting to know some electronics basics, interacting with the BeagleBone Black, and understanding the Linux operating system. Now we look under the hood of some of the interactions that have been taken care of for us by libraries, such as the Adafruit_BBIO Python library. In this chapter, we take a peek at the more complicated structure under the hood to gain a better appreciation for the work that libraries do for us and understand the hardware a little better, including the Cape expansion board system.

Linux Hardware Through The File System

One of the most important facts to remember is that everything in Linux is a file. This may seem like hyperbole, but everything is truly treated as a file. To aid in your understanding, you should first learn some low-level file commands and command-line tools that will help in your journey.

First, maybe the most important helper in learning about the Linux/Unix command-line environment is the `man` command. The `man` command is short for *manual* and provides a manual you can read through for most standard commands. In fact, you can even pull up `man` in formation on `man`:

```
root@beaglebone:# man man
MAN(1)                     Manual pager utils                    MAN(1)

NAME
       man - an interface to the on-line reference manuals

SYNOPSIS
       man [-C  file] [-d] [-D] [--warnings[=warnings]] [-R encoding]
       [-L locale] [-m system[,...]] [-M path] [-S list] [-e
       extension] [-i|-I] [--regex|--wildcard] [--names-only] [-a]
       [-u] [--no-subpages] [-P pager] [-r prompt] [-7] [-E encoding]
       [--no-hyphenation] [--no-justification] [-p string] [-t]
       [-T[device]]  [-H[browser]] [-X[dpi]] [-Z]
       [[section] page ...] ...
       man -k [apropos options] regexp ...
```

```
        man -K [-w|-W] [-S list] [-i|-I] [--regex] [section] term ...
        man -f [whatis options] page ...
        man -l [-C file] [-d] [-D] [--warnings[=warnings]] [-R
        encoding]  [-L locale] [-P pager] [-r  prompt] [-7] [-E
        encoding] [-p string] [-t] [-T[device]] [-H[browser]]
        [-X[dpi]] [-Z] file ...
        man -w|-W [-C file] [-d] [-D] page ...
        man -c [-C file] [-d] [-D] page ...
        man [-hV]

DESCRIPTION
 Manual page man(1) line 1 (press h for help or q to quit)
```

When the man page comes up, the command line will disappear and you find yourself looking at something more like a document that you can browse. Here are some useful commands in the man environment:

- **[space]**—Move one page down
- **[down arrow] or [j]**—Move one line down
- **[up arrow] or [k]**—Move one line up
- **[g]** - —Move to the top of the page
- **[G]** - —Move to the bottom of the page
- **[q]**—Quit

A man page generally starts with a list of all the argument and flag options. There are a number of options in the man example, so we can look at the man page for another important command, echo:

```
ECHO(1)                     User Commands                     ECHO(1)


NAME
       echo - display a line of text


SYNOPSIS
       echo [SHORT-OPTION]... [STRING]...
       echo LONG-OPTION


DESCRIPTION
       Echo the STRING(s) to standard output.


       -n     do not output the trailing newline


       -e     enable interpretation of backslash escapes


       -E     disable interpretation of backslash escapes (default)
```

```
        --help display this help and exit

        --version
                output version information and exit

 Manual page echo(1) line 1 (press h for help or q to quit)
```

The echo command is a very simple command that repeats the input string to the output with some very simple options. For example, using echo with -n will prevent the inclusion of a newline character after the string argument, whereas using -e and -E toggles the interpretations of backslash-escaped characters such as the newline (\n) character. Backslashes provide a way to include a normally nonprintable character in a string.

Pretty simple, but it doesn't show how we might start using the command. What if we wanted to write out "Hello, World!" to a file? We can combine the echo command with something called a redirect (>). The > character character>takes the output of the command line to the left and puts it into the file on the right. Here is an example where we look at a directory, run the command, and see the file has been added:

```
root@beaglebone:# ls
bbb-primer
root@beaglebone:# echo Hello, World! > new_file.txt
root@beaglebone:# ls
bbb-primer  new_file.txt
```

In this example, you can see a file, new_file.txt, has been created.

This is a good time to learn about another useful command: cat. The cat command takes a file or set of files as input, and prints their contents to the terminal. Here's how to check the contents of new_file.txt, for example:

```
root@beaglebone:# cat new_file.txt
Hello, World!
```

The more command is similar to the cat command. If the file is big and will scroll past the length of the terminal, you can use the more command to scroll page by page, similar to the man command.

Two more important concepts are the pipe (represented by |) and grep. The grep command allows us to search through the output of a file for a specific keyword, and the pipe command allows us to push the output of one command through the pipe and into the input of another command. These are powerful commands when you're hunting down information and/or processing data through a pipeline. For example, you can use another command calls ps, which can provide a snapshot of system processes. Using the aux argument to ps can produce a large amount of output. However, let's say you are only looking for processes running under the Apache web server that use the username www-data.

You would want to list the processes with the ps command but filter, using grep, on the lines that contain the string www-data, as shown in the following example:

```
root@beaglebone:# ps aux | grep www-data
www-data    875  0.0  0.3   5644  1996 ?        S    16:19   0:00 /usr/sbin/
apache2 -k
www-data    880  0.0  0.4 227056  2364 ?        Sl   16:19   0:00 /usr/sbin/
apache2 -k
www-data    881  0.0  0.4 227048  2360 ?        Sl   16:19   0:00 /usr/sbin/
apache2 -k
root       2712  0.0  0.1   1576   596 pts/0    S+   18:01   0:00 grep www-data
```

These are some of the core commands for quickly and easily manipulating files and checking file information in Linux. Because everything is a file in Linux, this forms the core of a powerful set of tools. Again, literally everything is a file. You might be tempted to think that the pins and hardware aren't files, but to Linux they are files.

Hardware in the File System

In previous code examples, we have turned on an LED with libraries that do the heavy work for us. Let's peek under the covers and turn on an LED attached to P9_11. Where do we even begin? We have to find the right file and understand how the files are based on blocks of memory that store the GPIO information. On the BeagleBone Black are four banks of GPIO information, and each bank holds information for 32 pins. These are, of course, indexed from zero. This gives us 128 possible GPIO ports (4×32). Only some of the pins are brought out to the headers. Only 67 of the 92 pins are available for GPIO; the remaining are for power, ground, and the analog inputs. Also, the pin positions don't relate to specific pin positions, so we need a mapping. In an attempt to make things easier, Table 8.1 and 8.2 show this mapping to P8 and P9, respectively.

TABLE 8.1 Mapping GPIO P8 to the GPIO Memory Locations

GPIO ID	Offset	Bank	Function	Pin	Pin	Function	Bank	Offset	GPIO ID
			DGND	1	2	DGND			
			VDD 3.3	3	4	VDD 3.3			
			VDD 5V	5	6	VDD 5V			
			SYS 5V	7	8	SYS 5V			
			PWR_BUT	9	10	SYS_ RESETN			
gpio30	30	0	GPIO	11	12	GPIO	1	28	gpio60
gpio31	31	0	GPIO	13	14	GPIO	1	18	gpio50

GPIO ID	Offset	Bank	Function	Pin	Pin	Function	Bank	Offset	GPIO ID
gpio48	16	1	GPIO	15	16	GPIO	1	19	gpio51
gpio5	5	0	GPIO	17	18	GPIO	0	4	gpio4
gpio13	13	0	GPIO	19	20	GPIO	0	12	gpio12
gpio3	3	0	GPIO	21	22	GPIO	0	2	gpio2
gpio49	17	1	GPIO	23	24	GPIO	0	15	gpio15
gpio117	21	3	GPIO	25	26	GPIO	0	14	gpio14
gpio115	19	3	GPIO	27	28	GPIO	3	17	gpio113
gpio111	15	3	GPIO	29	30	GPIO	3	16	gpio112
gpio110	14	3	GPIO	31	32	VDD_ADC			
			AIN4	33	34	GND_ADC			
			AIN6	35	36	AIN5			
			AIN2	37	38	AIN3			
			AIN0	39	40	AIN1			
gpio20	20	0	GPIO	41	42	GPIO	0	7	gpio7
gpio116	20	3					3	18	gpio114
			DGND	43	44	DGND			
			DGND	45	46	DGND			

TABLE 8.2 Mapping GPIO P9 to the GPIO Memory Locations

GPIO ID	Offset	Bank	Function	Pin	Pin	Function	Bank	Offset	GPIO ID
			DGND	1	2	DGND			
gpio38	6	1	GPIO	3	4	GPIO	1	7	gpio39
gpio34	2	1	GPIO	5	6	GPIO	1	3	gpio35
gpio66	2	2	GPIO	7	8	GPIO	2	3	gpio67
gpio69	5	2	GPIO	9	10	GPIO	2	4	gpio68
gpio45	13	1	GPIO	11	12	GPIO	1	12	gpio44
gpio23	23	0	GPIO	13	14	GPIO	0	26	gpio26
gpio47	15	1	GPIO	15	16	GPIO	1	14	gpio46
gpio27	27	0	GPIO	17	18	GPIO	2	1	gpio65
gpio22	22	0	GPIO	19	20	GPIO	1	31	gpio63
gpio62	30	1	GPIO	21	22	GPIO	1	5	gpio37
gpio36	4	1	GPIO	23	24	GPIO	1	1	gpio33

GPIO ID	Offset	Bank	Function	Pin	Pin	Function	Bank	Offset	GPIO ID
gpio32	0	1	GPIO	25	26	GPIO	1	29	gpio61
gpio86	22	2	GPIO	27	28	GPIO	2	24	gpio88
gpio87	23	2	GPIO	29	30	GPIO	2	25	gpio89
gpio10	10	0	GPIO	31	32	GPIO	0	11	gpio11
gpio9	9	0	GPIO	33	34	GPIO	2	17	gpio81
gpio8	8	0	GPIO	35	36	GPIO	2	16	gpio80
gpio78	14	2	GPIO	37	38	GPIO	2	15	gpio79
gpio76	12	2	GPIO	39	40	GPIO	2	13	gpio77
gpio74	10	2	GPIO	41	42	GPIO	2	11	gpio75
gpio72	8	2	GPIO	43	44	GPIO	2	9	gpio73
gpio70	6	2	GPIO	45	46	GPIO	2	7	gpio71

This is important information to have if you are to interact with the ports at the basic command-line level. Basic GPIO functionality is located in the `/sys/class/gpio/` directory. If you take a look at this directory, you can see which GPIO pins have been allocated:

```
root@beaglebone:# ls /sys/class/gpio/
export  gpiochip0  gpiochip32  gpiochip64  gpiochip96  unexport
```

You haven't explicitly allocated any of the pins yet, but you do have the basics you need to get started. The four `gpiochip` files relate to the four banks discussed before, and the `export` and `unexport` files provide special functionality. If you write a number associated with a GPIO memory location, that pin will be made available and configured. For turning on the LED, as discussed previously, you want to use pin P9_11. Checking Table 8.2, you see that this pin is attached to the GPIO ID 30, or offset 30 of bank 0. To export the pin, you just have to write a 30 to the export file:

```
root@beaglebone:# echo 30 > /sys/class/gpio/export
root@beaglebone:# ls /sys/class/gpio/
export  gpio30  gpiochip0  gpiochip32  gpiochip64  gpiochip96  unexport
```

We now have a directory for gpio30! The directory was created with some \ additional functionality.

```
root@beaglebone:# ls /sys/class/gpio/gpio30
active_low  direction  edge  power  subsystem  uevent  value
```

Working with these files provides the functionality used in the libraries. You can use the `cat` command to read the current values of the GPIO parameters, like so:

```
root@beaglebone:# cat /sys/class/gpio/gpio30/active_low
0
```

```
root@beaglebone:# cat /sys/class/gpio/gpio30/direction
in
root@beaglebone:# cat /sys/class/gpio/gpio30/edge
none
root@beaglebone:# cat /sys/class/gpio/gpio30/uevent
root@beaglebone:# cat /sys/class/gpio/gpio30/value
1
```

You can see that the pin is not configured for active low (zero equals false). The pin is set for input, it is not set up to watch for a voltage-change edge, uevent has no values, and the value being read at the pin is high, or 1. Reading these files is all it takes to understand the configuration on the GPIO pins. You can ignore the power and subsystem directories for now.

All you want to do is turn on your LED, so simply set direction to high. This is an easy way of saying you are going to an output and the value must be logic high. This turns on the LED. After the LED is on, you set the pin back to low and then you will get rid of the pin allocation by using unexport:

```
root@beaglebone:# echo high > /sys/class/gpio/gpio30/direction
root@beaglebone:# cat /sys/class/gpio/gpio30/direction
out
root@beaglebone:# cat /sys/class/gpio/gpio30/value
1
root@beaglebone:# echo low > /sys/class/gpio/gpio30/direction
root@beaglebone:# cat /sys/class/gpio/gpio30/value
0
root@beaglebone:# echo 30 > /sys/class/gpio/unexport
root@beaglebone:# ls /sys/class/gpio
export  gpiochip0  gpiochip32  gpiochip64  gpiochip96  unexport
```

As you can see, at the end of this sequence, the gpio30 pin has been exported and is no longer available. When you use a library such as the Adafruit_BBIO Python library, all of this is done for you. Although the Adafruit_BBIO library is a Python library, it is actually written in the C language, which is why we get very fast results, even in the sometime slower Python environment. In that C code you find many calls to functions that interact with the file system, abstracting away all of those interactions into a clean interface.

One Pin, Multiple Functions

Now we will get into some territory where things become a little more confusing. We know that for any given pin there can be multiple functions. This is controlled by something called a *mux*, short for *multiplexer*. The easiest way to think of the whole mux situation is that there are different parts of the hardware in the CPU that take care of different activities. Each pin can have up to eight different states (0 through 7) controlled by the mux. When you select a setting on the mux, the pin is connected, through the mux, to that part of the chip. Figure 8.1 attempts to clarify this a little by looking just at P9_11 (GPIO 30).

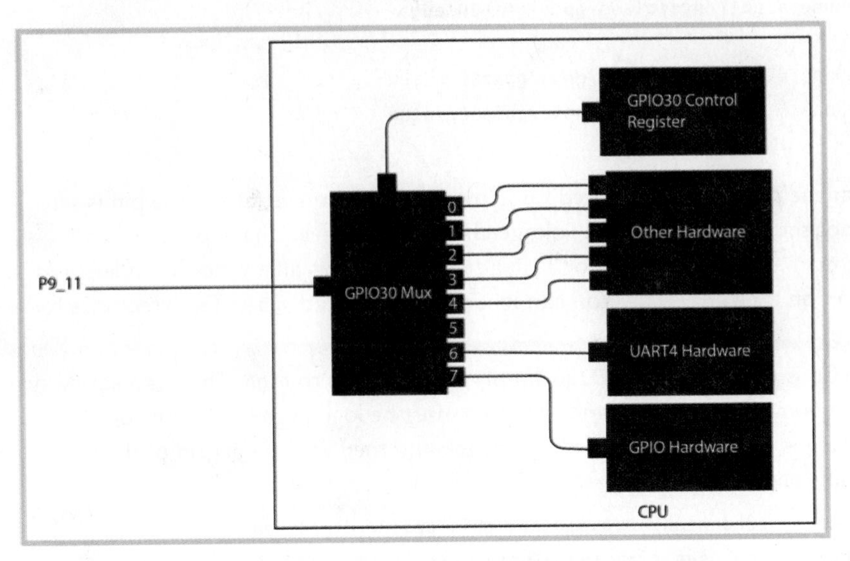

FIGURE 8.1 The multiplexed P9_11, or GPIO30.

Actually, there are a number of parameters for each individual GPIO and the related pin. To view the current status of all the GPIO, you can cat a file and, in this case, just look at GPIO30:

```
root@beaglebone:/bbb-primer/chapter08# cat \
/sys/kernel/debug/pinctrl/44e10800.pinmux/pins | \
grep "pin 30"

pin 30 (44e10878) 00000037 pinctrl-single
```

The important information for us in this line is the four-byte hex number, 0x00000037. This is the control register in the CPU associated with this GPIO. A *register* is a section of memory that is generally faster to access and used for control of the system. This register is bitmapped, meaning the different bits can have different meanings. The details are in the Sitara processor manual, a 4,966-page volume (as of revision K) that you don't need to dig through. Table 8.3 includes the relevant information.

TABLE 8.3 Register Definition for GPIO Control

Bit	Field	Description
31–20	Reserved	Nothing to look at here.
19–7	Reserved	Nothing to look at here either.

Bit	Field	Description
6	Slew Control	Select between faster or slower slew rate. 0: Fast 1: Slow
5	Receiver Active	Input enable value for the PAD. 0: Receiver Disabled 1: Receiver Enabled
4	Pull-Up/Down Type	Pad pull-up/pull-down type selection. 0: Pull-down selected 1: Pull-up selected
3	Pull-Up/Down Enabled	Pad pull-up/pull-down enable. 0: Pull-up/pull-down enabled 1: Pull-up/pull-down disabled
2–0	Mux Mode	Pad functional signal mux select.

Looking at this table, you know that it is a 32-bit register (bits 0 through 31) and you can ignore everything except the six least-significant bits. The slew rate allows the user to control how quickly a pin change happens. The slew rate is a setting for more advanced users. Receiver active defines whether we are input mode or output mode. The two bits surrounding the pull-up and pull-down resistors are interesting. We will talk more about pull-up and pull-down resistors in circuits in Chapter 9, "Interacting with Your World, Part 1: Sensors," but there is also the option for setting them up in the hardware.

Finally, the last three bits are the mux mode. The mux mode for all of the pins for GPIO is 7 (0b111), and the pins have varying functions associated with mux modes. Table 8.4 shows the default mode for each of the GPIO ports that are broken out to pins on the BeagleBone Black. This is generated directly from the kernel `pinctrl` call shown earlier.

TABLE 8.4 Default State of the Available BeagleBone Black Pins

GPIO	Physical Pin	Slew Control	Receiver	Pull Up/ Down	Pull Up/ Down Enable	Mux Mode
2	P9_22	Fast	Disabled	Pull Up	Enabled	1
3	P9_21	Fast	Disabled	Pull Up	Enabled	1
4	P9_18	Fast	Disabled	Pull Up	Enabled	1
5	P9_17	Fast	Disabled	Pull Up	Enabled	1
7	P9_42	Fast	Disabled	Pull Up	Enabled	1

GPIO	Physical Pin	Slew Control	Receiver	Pull Up/ Down	Pull Up/ Down Enable	Mux Mode
8	P8_35	Fast	Disabled	Pull Down	Enabled	7
9	P8_33	Fast	Disabled	Pull Down	Enabled	7
10	P8_31	Fast	Disabled	Pull Down	Enabled	7
11	P8_32	Fast	Disabled	Pull Down	Enabled	7
12	P9_20	Fast	Disabled	Pull Down	Enabled	7
13	P9_19	Fast	Disabled	Pull Down	Enabled	7
14	P9_26	Fast	Disabled	Pull Down	Enabled	7
15	P9_24	Fast	Disabled	Pull Down	Enabled	7
20	P9_41	Fast	Disabled	Pull Up	Enabled	7
22	P8_19	Fast	Disabled	Pull Up	Enabled	7
23	P8_13	Fast	Disabled	Pull Down	Enabled	7
26	P8_14	Fast	Disabled	Pull Down	Enabled	7
27	P8_17	Fast	Disabled	Pull Down	Enabled	7
30	P9_11	Fast	Disabled	Pull Up	Enabled	7
31	P9_13	Fast	Disabled	Pull Up	Enabled	7
32	P8_25	Fast	Disabled	Pull Up	Enabled	2
33	P8_24	Fast	Disabled	Pull Up	Enabled	2
34	P8_5	Fast	Disabled	Pull Up	Enabled	7
35	P8_6	Fast	Disabled	Pull Down	Enabled	7
36	P8_23	Fast	Disabled	Pull Up	Enabled	7
37	P8_22	Fast	Disabled	Pull Up	Enabled	7
38	P8_3	Fast	Disabled	Pull Up	Enabled	7
39	P8_4	Fast	Disabled	Pull Up	Enabled	7
44	P8_12	Fast	Disabled	Pull Down	Disabled	0
45	P8_11	Fast	Disabled	Pull Down	Disabled	0
46	P8_16	Fast	Disabled	Pull Down	Disabled	0
47	P8_15	Fast	Disabled	Pull Down	Disabled	0
48	P9_15	Fast	Disabled	Pull Down	Disabled	0
49	P9_23	Fast	Disabled	Pull Down	Disabled	0

GPIO	Physical Pin	Slew Control	Receiver	Pull Up/ Down	Pull Up/ Down Enable	Mux Mode
50	P9_14	Fast	Disabled	Pull Down	Disabled	0
51	P9_16	Fast	Disabled	Pull Down	Disabled	0
60	P9_12	Fast	Disabled	Pull Up	Enabled	0
61	P8_26	Fast	Disabled	Pull Up	Enabled	0
62	P8_21	Fast	Disabled	Pull Up	Enabled	0
63	P8_20	Fast	Disabled	Pull Up	Enabled	0
65	P8_18	Fast	Disabled	Pull Up	Enabled	0
66	P8_7	Fast	Disabled	Pull Down	Enabled	7
67	P8_8	Fast	Disabled	Pull Down	Enabled	7
68	P8_10	Fast	Disabled	Pull Down	Enabled	0
69	P8_9	Fast	Disabled	Pull Down	Enabled	0
70	P8_45	Fast	Disabled	Pull Down	Enabled	0
71	P8_46	Fast	Disabled	Pull Down	Enabled	0
72	P8_43	Fast	Disabled	Pull Down	Enabled	0
73	P8_44	Fast	Disabled	Pull Down	Enabled	0
74	P8_41	Fast	Disabled	Pull Down	Enabled	0
75	P8_42	Fast	Disabled	Pull Down	Enabled	0
76	P8_39	Fast	Disabled	Pull Down	Enabled	0
77	P8_40	Fast	Disabled	Pull Down	Enabled	0
78	P8_37	Fast	Disabled	Pull Down	Enabled	0
79	P8_38	Fast	Disabled	Pull Down	Enabled	0
80	P8_36	Fast	Disabled	Pull Down	Enabled	0
81	P8_34	Fast	Disabled	Pull Down	Enabled	7
86	P8_27	Slow	Enabled	Pull Down	Enabled	2
87	P8_29	Slow	Enabled	Pull Down	Enabled	2
88	P8_28	Fast	Disabled	Pull Down	Disabled	7
89	P8_30	Fast	Disabled	Pull Down	Enabled	7
110	P9_31	Fast	Disabled	Pull Up	Enabled	0
111	P9_29	Fast	Disabled	Pull Down	Disabled	0

GPIO	Physical Pin	Slew Control	Receiver	Pull Up/ Down	Pull Up/ Down Enable	Mux Mode
112	P9_30	Fast	Disabled	Pull Up	Enabled	0
113	P9_28	Fast	Disabled	Pull Down	Disabled	0
114	P9_42	Fast	Disabled	Pull Down	Disabled	0
115	P9_27	Fast	Disabled	Pull Down	Disabled	0
116	P9_41	Fast	Disabled	Pull Up	Enabled	0
117	P9_25	Fast	Disabled	Pull Up	Enabled	0

The most important information you can pull from this table as you are learning about the BeagleBone Black is that not all of the pin muxes are set to GPIO (7). This is due to the fact that the default boot for the BeagleBone Black has many pins already called into service for the HDMI, eMMC, and a serial protocol, different from the UART you learned about before, which was called SPI.

Hardware Configuration

How are all these configurations accomplished at the most basic levels and even at the startup of the board? The answer is in something called the Device Tree and the Device Tree overlay. The Device Tree ecosystem provides a way to make simple changes to the hardware configurations without recompiling the kernel. The Adafruit_BBIO library makes use of these overlays within the library. It was improved even more with the advent of something called the CapeManager for the BeagleBone Black—which brings us to Capes.

What is a Cape? A *Cape* is just a hardware extension of the BeagleBone Black itself. It plugs in to the top of the board with matching pins. A Cape defines the hardware it needs or utilizes by additions to the Device Tree and generally adds some new capability. This is the most basic description one can have of a Cape, because from there a Cape is defined by the imagination and need of the designer. An entire ecosystem of Capes exists to provide anywhere from almost no functionality to expansive functionality. An incredible basic Cape, the SparkFun ProtoCape, is shown in Figure 8.2.

FIGURE 8.2 The SparkFun ProtoCape stacked on a BeagleBone Black, compared with a bare BeagleBone Black.

The ProtoCape has a basic role: It brings all of the pins to the electronics equivalent of a blank canvas for a developer to wire and test a new design. This is incredibly useful as your skill base includes more prototyping and development, or if you have a one-off project for your home that you want to wire up more permanently than on a breadboard.

Another great Cape from SparkFun is the CryptoCape, shown in Figure 8.3.

FIGURE 8.3 The SparkFun CryptoCape.

The Crypto Cape may look a lot like the ProtoCape with a few more components due to the fact that it has a large prototyping area of its own. An important aspect of the CryptoCape in comparison to other Cape boards is that the headers are not brought out to the top. However, the Crypto Cape provides some very important hardware-based utilities, including the following:

- A real-time clock enables the board to keep its own, more accurate time and not have to rely on time servers. If you look at the CryptoCape in Figure 8.3, you see there is a spot for a battery with the real-time clock in the lower left of the board. This ensures that time is kept between power cycles. A handy utility that takes very little power to maintain. Therefore, the battery will last a long time.
- A trusted platform module and other cryptographic chips allow for trusted hardware encryption and decryption without relying on the operating system. These are often intensive procedures that can use up a lot of operating system time. Moving this functionality off to hardware speeds up the process considerably and frees up the operating system.
- An Atmel ATMega328, which essentially provides a small Arduino to which you can offload other hardware and timing-specific functions such as Pulse Width Modulation (PWM, which is discussed more in Chapter 11, "Interacting with Your World, Part 2: Feedback and Actuators") and watching for a pin change.

A much more advanced pair of Capes available from Element14 are the BB-View LCD Capes. These Capes provide an LCD screen in either a 4.3-inch or 7-inch format, and the screen is touch sensitive and provides feedback to the BeagleBone Black as an input device. This is immensely useful in projects and opens up a world of possibilities. It works with your finger or a generic stylus. The stylus shown with the 4.3-inch BB-View cape in Figure 8.4 is one from my son's Nintendo 3DS. Thanks, Sean!

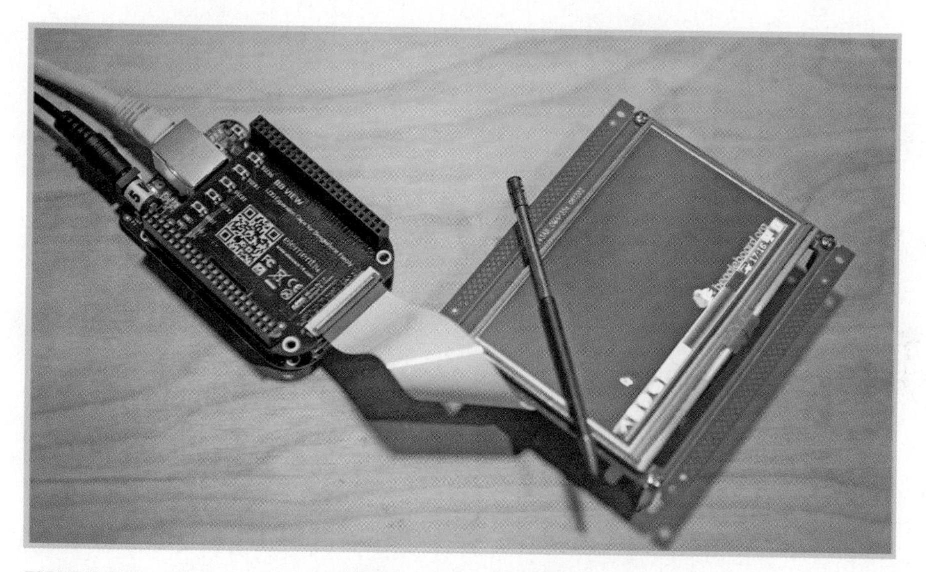

FIGURE 8.4 The Element14 4.3" BB-View Cape.

Now the two SparkFun Capes shown are, in many way, very basic. Obviously the ProtoCape provides no functionality other than access to a EEPROM that stores configuration information in the Cape world. It is a blank canvas. Many of the functions of the CryptoCape are also a part of Linux. With greater power, sometimes, comes greater complexity. At the time of writing, a number of patches need to be loaded to the BeagleBone Black for the Element14 BB-View devices. The BB-View Cape also turns off the network by default requiring you to connect from the USB interface and update the file `/etc/network/interfaces` for eth0 to turn networking back on.

As discussed before, you must read the documentation. (This becomes increasing crucial as we move forward.) The datasheets are often available online, and Element14 and SparkFun both provide wonderful guides to get us going.

Now that you understand the basics of hardware and some of the underlying configuration of the BeagleBone Black, it is time to work toward building up to bigger projects.

Interacting with Your World, Part 1: Sensors

Whereas some projects for embedded systems such as the BeagleBone Black simply enclose the board attached to the network and use the board as a computer resource or as a server, many projects such as robotics, home automation, and environmental sensing rely on gathering information about the physical world around the project. In this chapter, you will get a better feel for how to utilize sensors on the BeagleBone Black.

Sensor Basics

We should start with a basic concept: What is a sensor? A *sensor* is a form of transducer, which is a device that takes energy as input and outputs it as another form of energy. Sensors, in general, convert a change in some energy and produce an electrical signal as a result. In this chapter, you will learn the basics of sensors and how to properly read and interpret their signal. In general, you will encounter two types of sensor packages: extremely basic sensors where you are required to detect the change in the environment, and those that are more complex and someone else has already packaged them so you can access the values over a serial (or other) connection. We will start with the former and move toward the latter.

One of the most basic things you can sense is on or off. Very simple. You do it all the time. Flip on a light switch, you get light and you quickly sense the environment has changed. Push a button on the remote, and you see that the TV comes on. All a switch does is complete a circuit. Figure 9.1 shows a schematic for a simple button circuit on the BeagleBone Black.

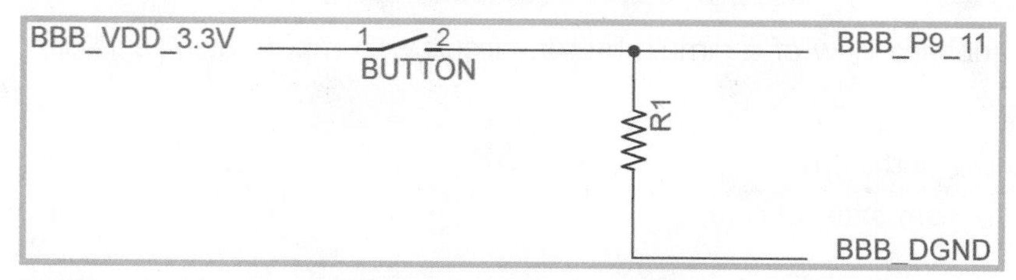

FIGURE 9.1 A simple button circuit with pull-down resistor.

Let's trace around what is happening in this circuit. Power is provided by the BeagleBone Black 3.3V source. One side of the button is connected to this voltage source. The other side of the button is connected to a GPIO pin on the BeagleBone Black and through a resistor to ground. The resistor to ground is an important concept. When the button is not pressed, there is no voltage connected from the 3.3V to the GPIO pin, so the GPIO pin will just float at a voltage. Just because the pin isn't driven to logic high doesn't mean it is sitting at logic low. It is, literally, floating and picking up voltage from electromagnetic waves, static electricity, and other noise sources. Tying the line to ground through a resistor in this manor ensures that the signal is at logic low when another voltage is not applied. This is called a *pull-down resistor*.

When the button in this circuit is pressed, the line connected to the GPIO pin is brought up to 3.3V. Again, pretty simple. The breadboard version of this is shown in Figure 9.2. What you need to do now is take some kind of action with the button press. For this, turn to the handy Adafruit Python libraries. Listing 9.1 shows a simple Python program to read a button.

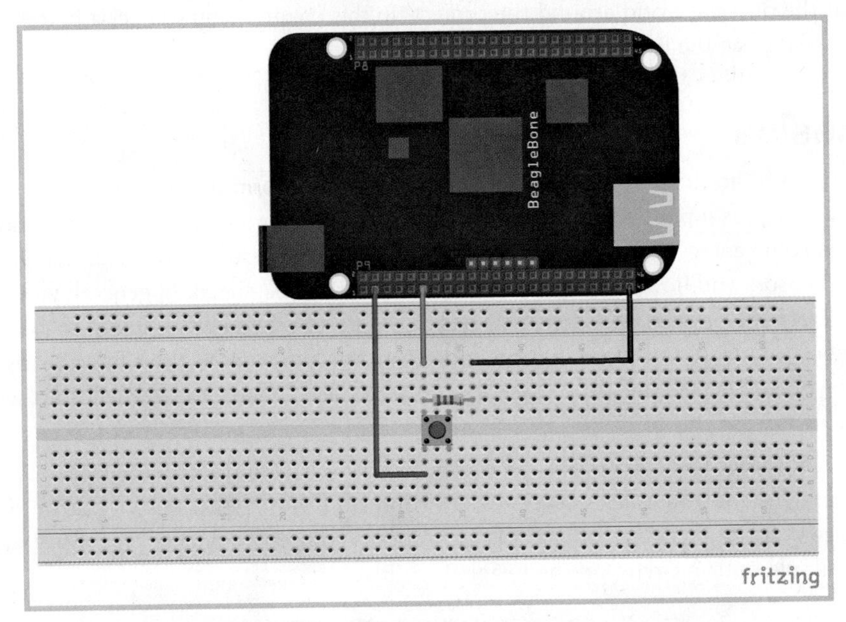

FIGURE 9.2 Breadboard view of a simple button circuit with a pull-down resistor.

LISTING 9.1 simple_button.py

```
01: import Adafruit_BBIO.GPIO as GPIO
02: import time
03:
04: # Define program constants
```

```
05: BUTTON_PIN = 'P9_11'
06: OFF        = 0
07: ON         = 1
08:
09: # Configure the GPIO pin and set the initial state of
10: # variables to track the state of the button.
11: GPIO.setup(BUTTON_PIN, GPIO.IN)
12: button_state_old = OFF
13: button_state_new = OFF
14:
15: # print out a nice message to let the user know how to quit.
16: print('Starting, press <control>-c to quit.\n')
17:
18: # Execute until a keyboard interrupt
19: while True:
20:     try:
21:         # Check the state of the pin. If it is
22:         # different than the last state,
23:         # print a message.
24:         button_state_new = GPIO.input(BUTTON_PIN)
25:         if button_state_new != button_state_old:
26:             if button_state_new == OFF:
27:                 print('Button transitioned from off to on.')
28:             else:
29:                 print('Button transitioned from on to off.')
30:
31:         # Update the stored button state and then wait a tenth of a second.
32:         button_state_old = button_state_new
33:         time.sleep(0.1)
34:
35:     except KeyboardInterrupt:
36:         GPIO.cleanup()
```

This is a simple program, and you should be familiar enough with Python and the Adafruit library to follow along. Everything up to line 13 sets up the program. Lines 23 through 32 execute in a loop where the state of the pin is read. If it has changed from the last time through the loop, the program prints a message based on the transition. The previous button state is updated with the new state, and the program sleeps for a short time to allow the computer to do other stuff. This same kind of circuit works for a switch and other types of simple on/off sensors. There could be reasons why you would want the circuit to work the opposite way—that is, sit at logic high when the button is not pressed and go low when it is pressed.

To accomplish this, tie the sensor line through a resistor to the 3.3V supply, just like you did with ground in the other scenario, and attach the other end of the button or switch to ground. The resistor is now called a "pull-up" resistor. Figure 9.3 shows the updated schematic.

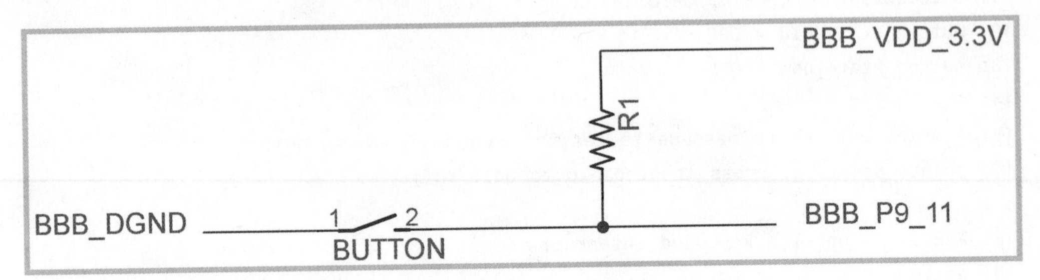

FIGURE 9.3 A simple button circuit where off is pulled up to 3.3V.

You've seen this pull-up scenario before, and it's used often. Remember how serial data lines sit at logic high until the start bit goes low? That is accomplished by tying the line to a pull-up resistor.

The code is simple enough, but polling for button presses like this is needlessly complex if all you literally need to do is wait for a single button press. The Adafruit GPIO library lets you just wait for an action, and holds the program from doing anything else until that action occurs, using a function called `wait_for_edge()`. Now keep in mind that when you are waiting for an edge with this function, your program can't do anything else. However, let's say that instead of wanting to just print a state change, you want to execute an external program and turn on an LED while the program is executing. In this case, the circuit becomes a little more complex, as shown in Figure 9.4, but the source code become a little easier to understand, as shown in Listing 9.2.

LISTING 9.2 not_as_simple_button.py

```
01: import Adafruit_BBIO.GPIO as GPIO
02: import time
03: import subprocess
04:
05: # Define program constants
06: BUTTON_PIN = 'P9_11'
07: LED_PIN    = 'P9_12'
08:
09: # Configure the GPIO pins and set the initial state of variables
10: # to track the state of the button.
11: GPIO.setup(BUTTON_PIN, GPIO.IN)
```

```
12: GPIO.setup(LED_PIN, GPIO.OUT)
13: GPIO.output(LED_PIN, GPIO.LOW)
14:
15: # print out a nice message to let the user know how to quit.
16: print('Starting, press <control>-c to quit.\n')
17:
18: # Execute until a keyboard interrupt
19: try:
20:        while True:
21:               # Wait for the BUTTON_PIN to have a falling edge,
22:               # indicating the button has been pressed.
23:             GPIO.wait_for_edge(BUTTON_PIN, GPIO.RISING)
24:
25:             # Button has been pressed so turn on the LED and start
26:             GPIO.output(LED_PIN, GPIO.HIGH)
27:             subprocess.call(['/path/to/the/program', '-argument'])
28:
29:             # Program done, turn off LED and start waiting again.
30:             GPIO.output(LED_PIN, GPIO.LOW)
31:
32: except KeyboardInterrupt:
33:     GPIO.cleanup()
```

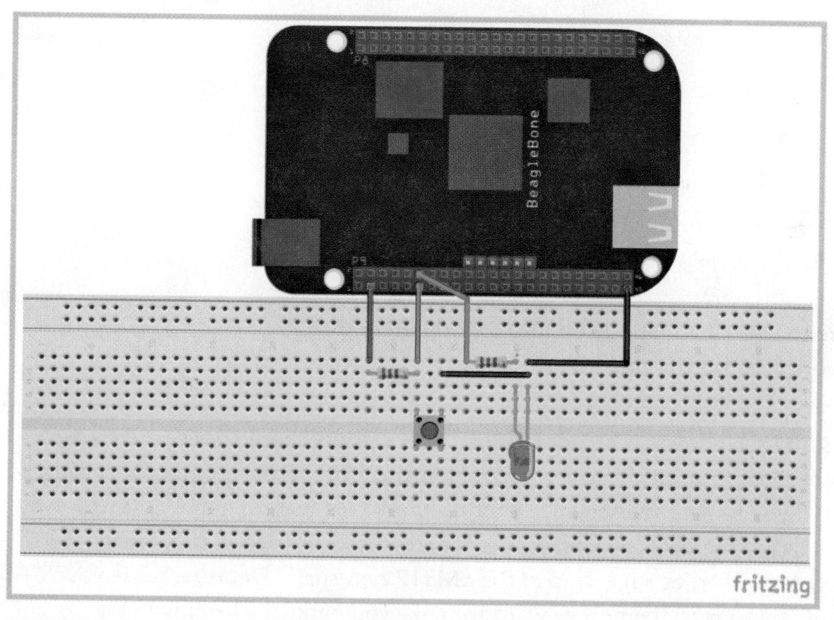

FIGURE 9.4 A simple pushbutton circuit with an LED indicator.

This program wouldn't generally run from the console. Instead, it would be set up to run in the background when the board is turned on and would respond on its own to the button presses to run the activity. Setting up a program to run in the background like this is something we cover in the next chapter, but you can already see how adding simple hardware sensing to a computing resource can make for a much more involved project without the project itself becoming overly complex.

As discussed previously, sensors transform one type of energy into a signal we can use. In the case of buttons and switches, the mechanical energy of the button change or switch change is translated into an electrical signal—simple and effective. This concept can scale up very quickly. Want to guess how some simple joysticks operate? Take a look at Figure 9.5 for a potential joystick circuit. If you put a lever with a cross-member at the bottom so that when the lever is moved in different directions, different buttons (or combinations of buttons) are pressed, you can sense not only a mechanical force, but the direction it went as well!

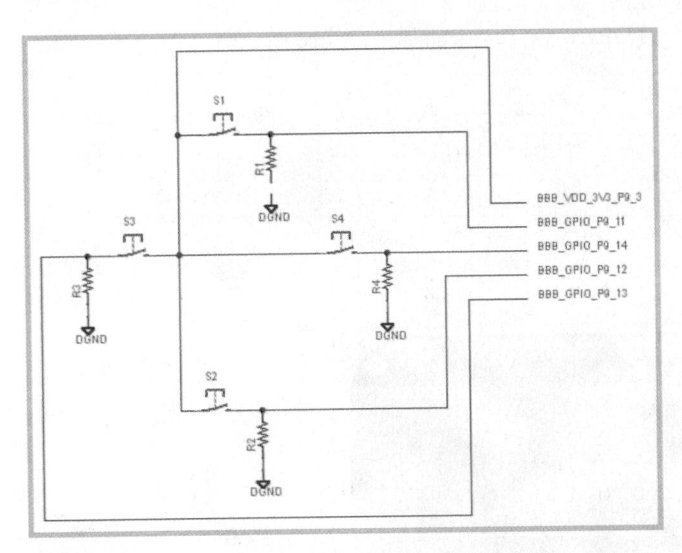

FIGURE 9.5 A simple joystick schematic.

You can also take this simple multibutton idea in a different direction. Want to guess how your keyboard works? So button presses are simple, but what about rotation? One way to sense rotation is with variable resistance.

Variable resistors and their cousin, the potentiometer, are pretty simple devices. They have an input and an output connected to sweeper arm. As the arm moves, the resistance across the wire changes. If you look at the schematic for the 3.3V source in Chapter 7, "Expanding the Hardware Horizon," there is a potentiometer/variable resistor used in that circuit to tune in the voltage regulator, as suggested in the LM317 datasheet. Datasheets are your friends. They want you to read them; they want to make you happy by helping make your project work. You know a resistor causes the voltage to change in proportion to the resistance encountered based on the formulas in Chapter 4, "Hardware Basics." Therefore, if you change the resistance, you can change the voltage that comes through the circuit.

Generally, for this kind of circuit, a potentiometer is used because it mimics a voltage divider. A voltage divider circuit, shown in Figure 9.6, provides a way of stepping down voltages for signals and lower-current lines. It should not be used to change the voltage on a supply line because the higher current utilization and complex resistance of a load can change the properties of the voltage divider in undesired ways. This is why I used a voltage regulator in Chapter 7. However, for something like a signal line, a voltage divider is very predictable. The voltage tapped between the two resistors is calculated with the following equation:

$$V_{out} = \frac{R_2}{R_1 + R_2} \times V_{in}$$

So, for example, to change a signal level from 5 V to 3.3 V we follow:

$$V_{out} = \frac{3.3k\Omega}{1.7k\Omega + 3.3k\Omega} \times 5V$$

$$V_{out} = \frac{3.3k\Omega}{5k\Omega} \times 5V$$

$$V_{out} = 0.66 \times 5V$$

$$V_{out} = 3.3V$$

With a potentiometer, as you turn the knob and sweep the arm, you change the proportional resistance. As an example, Table 9.1 shows the measured resistance at a couple points in the rotation of a potentiometer. The table shows the input voltage as 1.8V for a specific reason we will hit upon in just a moment.

TABLE 9.1 Potentiometer Values Through an Arm Sweep

Input	R1	R2	Output
1.80	1	10280	1.80
1.80	1710	8600	1.50
1.80	3110	7230	1.26
1.80	5280	5090	0.88
1.80	7030	3400	0.59
1.80	8600	1790	0.31
1.80	10270	0	0.00

What can you learn from this table? First, as the potentiometer turns, we observe that the values of R1 and R2 stay inversely proportionate. Any amount R1 drops, R2 increases. There are small variations in the table due to the limited resolution of the multimeter in use, but the relationship is clearly visible. The second thing to note is that when R1 is at a minimum, we are almost in a short-circuit scenario. The resistance was just 1 Ω, or less with measurement error. You will see that, sometimes, this calls for a resistor before the voltage divider to make sure that we never short-circuit. Figure 9.6 shows the schematics for a conceptual voltage divider and for a potentiometer circuit.

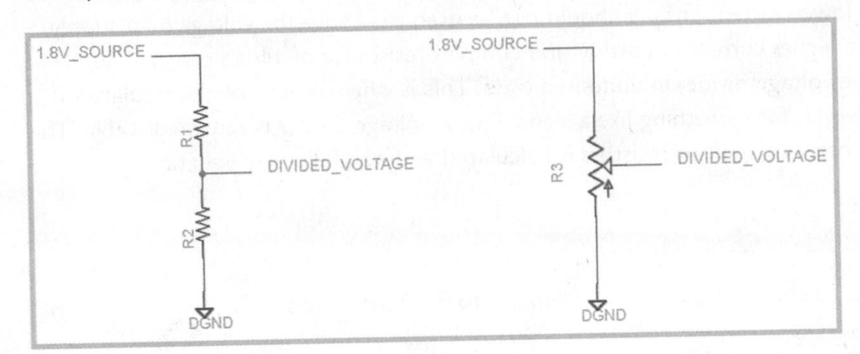

FIGURE 9.6 Schematics for a voltage divider and a potentiometer.

Analog Versus Digital

So, now you have a potentiometer that, when turned, provides different voltages. If you could measure the voltage, you could make a program on the BeagleBone Black have knowledge of its position. We are translating a relative position into a voltage, and this is where we get into the meat of a device that is used a lot for sensing: the analog-to-digital converter.

So what is the difference between analog and digital that you would need a converter? In the most generic over terms, digital is based on logic states—high and low; on and off. The binary positions trigger the different levels. Analog, on the other hand, can be anywhere in a range. A heartbeat is a very common analog signal you would recognize fairly easily. Figure 9.7 shows an oscilloscope capture of my wife's pulse. Notice that it doesn't just jump between two levels; instead, it is a smooth curve.

We will go into more depth on the concept of sampling a little later in the chapter, but the basic idea is simple. An analog-to-digital converter has a circuit inside it that, for a fraction of a second, holds a potentially changing voltage at one level and translates that into a number in a range. The analog-to-digital converter on the BeagleBone Black is a 12-bit converter. This means that when the input is at 0V, the value stored is 0b000000000000, and when the voltage is at the maximum for the converter, the value stored is 0b111111111111. What is the decimal equivalent? Here is a simple equation for how many values can be represented in a number of bits:

$$values = 2^n$$

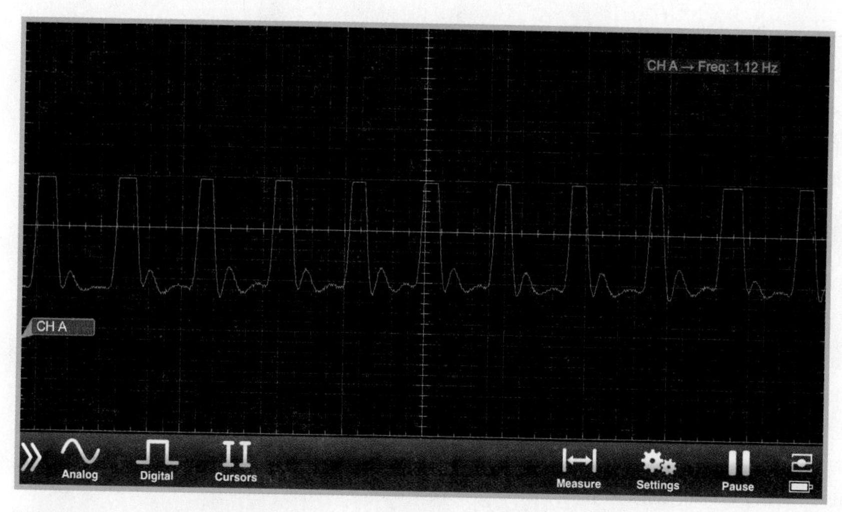

FIGURE 9.7 A capture of a heartbeat.

In this equation, *values* represents the quantity of numbers that can be represented and *n* is the number of bits. For a three-bit number, the maximum number of values is

$$values = 2^z$$
$$values = 8$$

So for three bits we have eight possible values, but the maximum value needs to take into account the lowest number, which is 0, thus leaving the maximum value as one less than *values*. So, the 12-bit analog-to-digital converter goes from 0 to

$$max = 2^n \ \ 1$$
$$max = 2^{12} - 1$$
$$max = 4095$$

This means that in our 12-bit analog-to-digital converter, a maximum reading gets is a value of 4095. The maximum voltage for the analog-to-digital converter on the BeagleBone Black is 1.8V. This is very important and bears emphasis:

The maximum voltage for the analog-to-digital converter on the BeagleBone Black is 1.8V.

Do not push direct 3.3V or 5V or anything else that is over 1.8V into the analog-to-digital convert pins or you will cause damage. Conveniently, the analog-to-digital pins have a nearby source that provides the 1.8V required. This is also why I used a 1.8V source earlier. Now, you'll want to know how that resolution translates, so you take the range of 0 to 1.8 and divide it up into 4,096 units, known as *counts*, like so:

$$1.8 \div 4095 = 0.0004$$

This means that a 0.0004V change will alter the digital value by one bit and will translate vice versa. It all seems easy enough. There is just one problem, though: noise. The analog-to-digital converter will do its best to capture a value, but other factors and noise sources on the board can vary the actual voltage level read. Running a quick test using the potentiometer sitting still at the halfway mark, I can see a variation of one or two bits. This means that the values of the first couple bits can be ignored. This isn't that bad, though. Because we know that one count is 0.0004V, and we can safely ignore the first two bits (a value of 3), this results in an uncertainty range of 0.0012V, which is just over one-thousandth of a volt.

Now that you have a better understanding of how you can use this set of pins to read a voltage, you can get a better understanding of sensor readings. Seven analog-to-digital pins are available on the BeagleBone Black, all on P9. There is also the previously mentioned analog 1.8V supply as well as a special analog ground pin that should be used for grounding of the sensor. Using this voltage and ground helps keep noise in the analog-to-digital conversion at a minimum.

A fun sensor for understanding the sampling of an analog signal is a pulse sensor. The SparkFun Pulse Sensor kit (SEN-11574) is simple to use and can be up and running rather quickly. Figure 9.7 shows a reading from an oscilloscope of this sensor with power supplied by the BeagleBone Black 3.3V output. Of course, you can't run the sensor directly to the analog-to-digital convert pins because of their 1.8V limit. Instead, connect the output through a voltage divider to bring the maximum voltage below the 1.8V threshold. Figure 9.8 shows this configuration.

The code for the collection of data from this sensor should seem trivial to you by now (see Listing 9.3).

LISTING 9.3 heartrate.py

```
01: import Adafruit_BBIO.ADC as ADC
02: import time
03:
04: # Define program constants
05: ANALOG_IN       = 'AIN0'
06: SAMPLE_RATE     = 100       # Hertz
07:
08: # Short function to handle a bug in the ADC drivers where the
09: # value needs to be read twice to get an actual value
10: def read_adc(adc_pin):
11:     ADC.read(adc_pin)
12:     return ADC.read(adc_pin)
13:
14: # Configure the ADC
15: ADC.setup()
```

```
16:
17: # Execute until a keyboard interrupt
18: try:
19:     while True:
20:         value = read_adc(ANALOG_IN)
21:         print(value * 1.8)
22:         time.sleep(1/SAMPLE_RATE)
23:
24: except KeyboardInterrupt:
25:     pass
26:
```

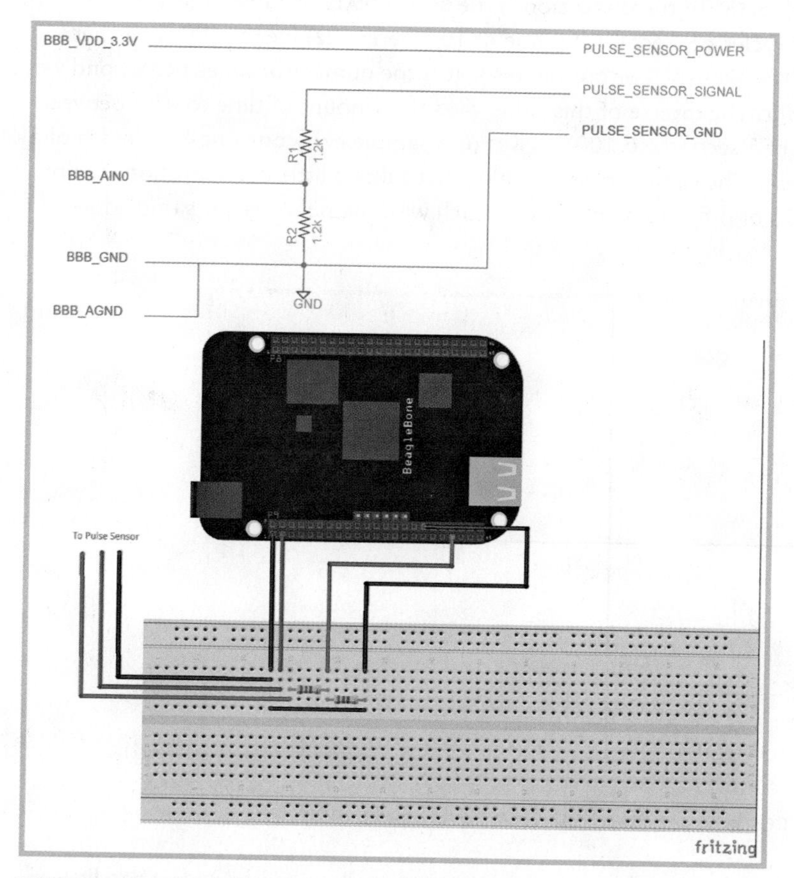

FIGURE 9.8 SparkFun pulse sensor connections.

The `read_adc` function simply reads the analog-to-digital converter twice, which is necessary at the time of writing due to a bug in the analog-to-digital drivers, as reported by Adafruit in the Adafruit_BBIO documentation. This is a great example of why documentation is your friend!

Running this program normally from the command line will stream many rows of numbers along the terminal. What we need to do is send the output from the program into a file rather than the screen. You use a command-line redirect for this:

```
root@beaglebone:/bbb-primer/chapter09# python heartrate.py > data.txt
```

This will send all the data out to a file that you can than pull into another program, such as Excel, for plotting.

Sample Rates

An important point of focus for the discussion is the SAMPLE_RATE variable. This value is measured in Hertz, or cycles per second. A value of 100 Hertz (Hz) means 100 times per second. What the 100 is can vary. For our interests, it is the number of times per second we will take a measurement. The inverse of this value gives the amount of time to wait between samples as a fraction of a second, so 100Hz gives us a sample every one one-hundredth of a second, or 0.01 seconds. The sample rate is a value that takes a little consideration. Figure 9.9 shows us four plots of different sample rates, each with the pulse sensor sampled at 100Hz behind it.

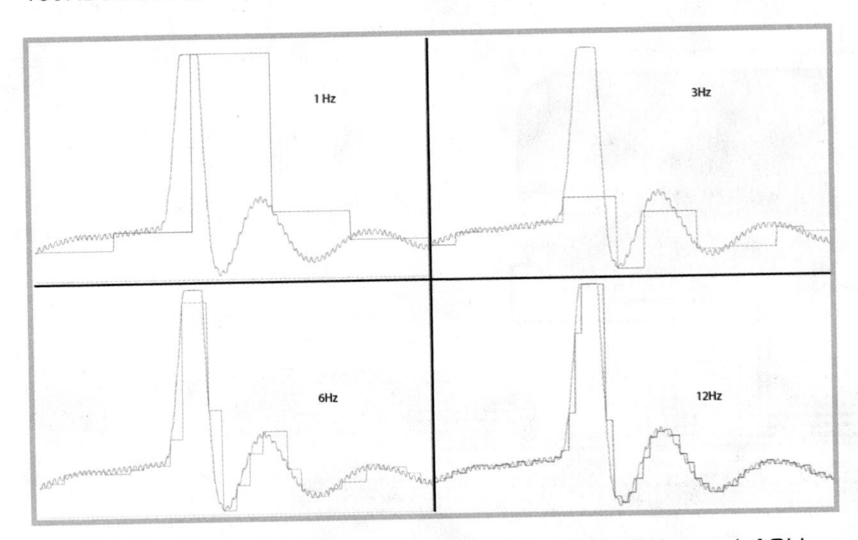

FIGURE 9.9 Pulse sensor sampled at 1Hz, 3Hz, 6Hz, and 12Hz.

If we sample too often, we collect far more data than we need, and if we collect too little data, we risk not capturing the waveform. At 100Hz, we have collected 500 data points for this one small section of a pulse. We could drop down to 1Hz, but then what happens? We reduce the amount of data stored to only five data points, but we can't possibly reconstruct the waveform that is so clear at 100Hz.

One thought would be to see how far apart the smallest time period between parts of the waveform falls and sample at that rate. A quick measurement of the valley just after the large peak to the next, smaller peak is about 0.3 seconds, so what happens if we use 3Hz? We are definitely closer to sampling the waveform, but if we look at the big peak when sampled at 3Hz, we didn't capture the magnitude of the wave even closely. This is because one sample happened to fall on the rise of the pulse and the next sample landed inside the trough following. The 3Hz and 1Hz examples are therefore under-sampled to properly reconstruct the waveform.

Now, we know we were closer with 3Hz, and we know that the smallest feature we need to capture is roughly related to the 3Hz at 0.3 seconds, so maybe we would be in better shape if we double the sample rate to 6Hz. The 6Hz sampled line shows a pretty good reconstruction of the waveform! It is rough, but it is a pretty accurate representation. Turns out this is a well-known feature in converting from continuous data, such as the smooth waveform from the sensor to sampled, discrete values. The minimal sample rate for an accurate reconstruction of a wave is twice the frequency of interest. This is where you'll hear Nyquist Sampling, Nyquist Frequency, and other terms involving the name Nyquist, who was one of the people who discovered this phenomena. The discovery actually relates back to how fast we need to sample a communications signal waveform in order to recover the signal being transmitted. The reasons for this are really a lot of fun if you like Calculus and Fourier Transforms, but we won't discuss these topics in this book.

So we know that twice the frequency of interest will allow us to reconstruct the basic waveform, but what if we want a little better sample to work from? A great rule-of-thumb is to use four times the frequency of interest to get a good wave reconstruction. You can see that in the 12Hz sample rate in Figure 9.9, we are much closer to the actual waveform, but we are not using anywhere near the amount of data for storage—only 42 data points at 12Hz as opposed to the 500 at 100Hz, and yet we can still reconstruct our waveform!

There is still more improvement, however, to be had. If you look at the 100Hz waveform alone, there is a constant ripple throughout the wave. This is noise in the analog-to-digital sampling. If we sampled the output of the potentiometer readings back in Table 9.1, we would see a similar bouncing. This is the lowest couple bits of the sample bouncing around. Remember that we calculated in the 12-bit analog-to-digital converter with a range of 0V to 1.8V that a one bit change in the least-significant bit is a change of 0.0004 volts. In the world of delicate scientific observations, systems have samplings at a much finer resolution than this, but those systems are enclosed, shielded, grounded, and use many other precautions to keep the noise to a minimum. This is an open development board sitting on a wooden work bench with all kinds of electrical noise floating around. A change in the sample of 0.0004V is nothing.

All told, we are seeing a noise ripple of about 0.003V. This includes all the noise sources in the system. How can we account for this noise so it isn't mistaken for data? There are a number of methods, but they all have one thing in common: a loss in waveform resolution. We have already looked at one method: reducing the sample rate. At 100Hz, we are sampling far faster than we need to for data we can capture just fine at 12Hz. As you saw in Figure 9.9, a 12Hz sample rate gives us a nice waveform with all the details we need.

Although there are reasons to sample faster than double the Nyquist and ways to get rid of that noise, we really should focus on sampling at a reasonable rate. As you saw, this reduces the amount of data we need to store, removes the need to worry about certain noise artifacts, and results in less processor effort because samples occur less frequently.

In the next chapter, we will build a project based on sensors to monitor the conditions in a room. We will also send the data out to storage on the Web and even provide the ability to plot the data!

Remote Monitoring and Data Collection

You've learned a lot of basics for interacting with the BeagleBone Black, and in Chapter 9, "Interacting with Your World, Part 1: Sensors," you learned how to capture data from the world around you. That may be the most interesting use of computers like the BeagleBone Black—capturing and interacting with the world around you. Making that information accessible is a next positive step in that direction. In this chapter, you will build a project to monitor some conditions in the environment around your board and publish that data to make it more accessible. You will even start plotting the data!

Project Outline

Let's take a moment and outline a project. You want to collect data on the environment around the BeagleBone Black and make that information easily accessible. Here are some goals:

- Select an environmental condition you find interesting.
- Collect the data at a reasonable rate.
- Publish the data to a storage space on the internet.
- Start data collection and publishing automatically when the BeagleBone Black is powered on.

So, what are some interesting environmental conditions in a workspace? Temperature is always popular. The workshop/office area in my house has a lot of computer gear and devices that run all the time, and it would be interesting to know the temperature in the room over time. We might also be interested in knowing if the lights are left on or off. We will go with these two basic measurements for now. With the skills you learned in previous chapters, it should be clear how to add other sensors to the project.

You need to line up the sensors you are going to use first, read up on their datasheets, and understand how to integrate them into the BeagleBone Black. For temperature, you will use a TMP36 device (SparkFun SEN-10988, Adafruit Prod# 165, Element14/Newark Part# 19M9015). This is an easy-to-use temperature sensor in a small package, shown in Figure 10.1. For light sensitivity, you will use something called a photocell (SparkFun SEN-09088, Adafruit Prod# 161).

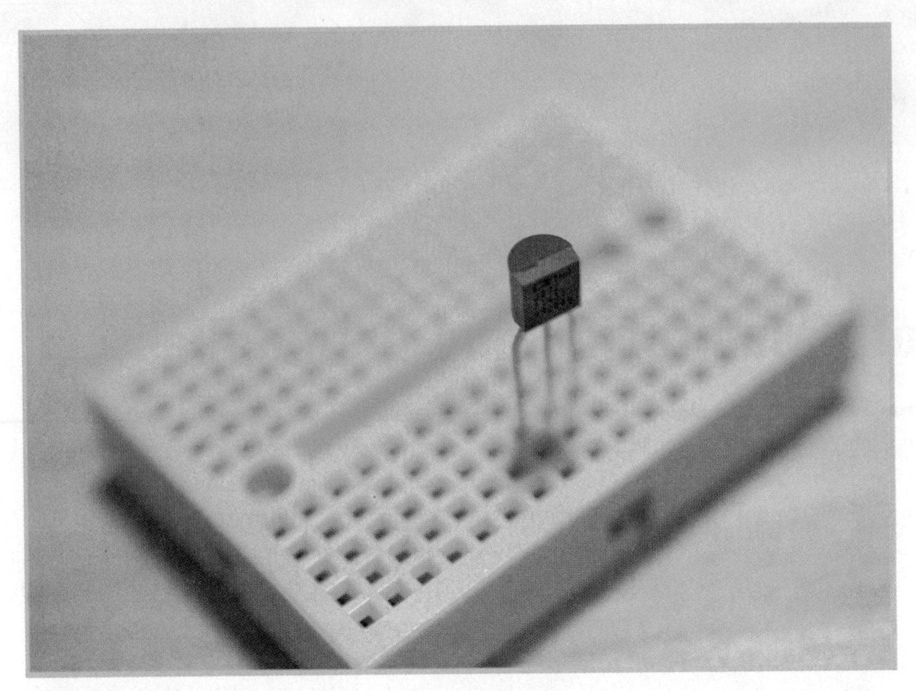

FIGURE 10.1 A TMP36 low-voltage temperature sensor.

This sensor has a very simple operation, with only three connections: a voltage in (also known as $Voltage_{supply}$ or V_s) that needs to be between 2.7V and 5.5V, a voltage out (V_{out}), and a ground connection. The current used by the device is less than 0.5 μA, so there is no problem powering it. The device outputs a voltage that is proportional to the temperature and works in the range of -40°C to 125°C (-40°F to 257°F). Hopefully, you don't have an indoor environment that exceeds those boundaries. From what you have already learned, you know you will need to use an analog-to-digital input to read the voltage and translate that into a temperature reading. You can't use the 1.8 V analog-to-digital supply, so you will need to consider using the 3.3V supply or 5V supply. No matter which voltage input you select, the output is 0V to 2V over the stated temperature range. Every 10mV in the output is 1°C. The 2V maximum is over our 1.8V maximum. However, if you look at the output voltage versus temperature chart in the datasheet (reproduced in Figure 10.2), you see that over the operational range the voltage stays below 1.8V.

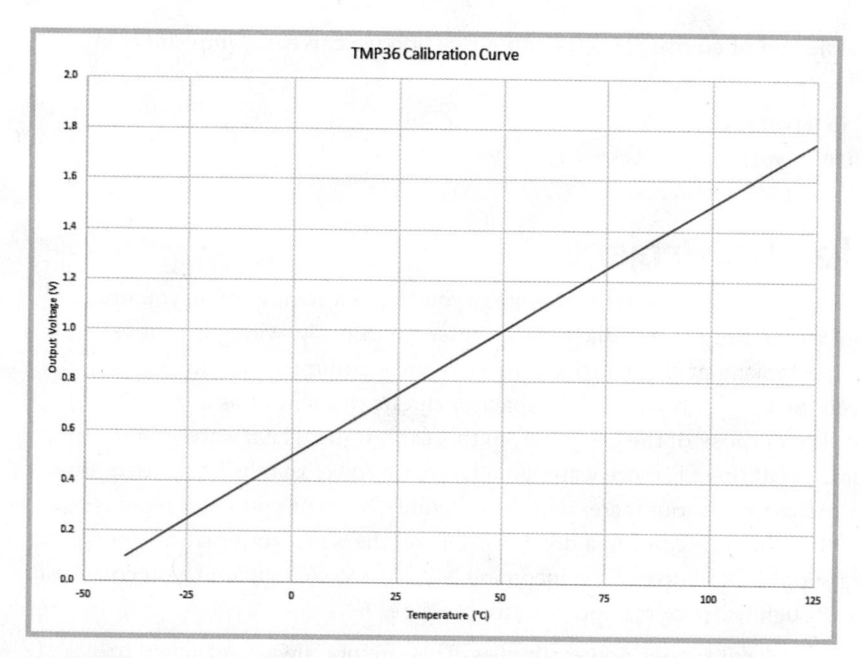

FIGURE 10.2 The TMP36 output voltage versus temperature.

The datasheet documents the curve as being extremely linear, with a slope of 0.10 V/°C and a calibrated point of 0.750V at a temperature of 25°C. You can find the simple equation of the line for our calibration then by finding the y-intercept. In this case, the y-intercept is the point voltage output at 0°C:

$$y = m \cdot x + b$$
$$0.750V = \left(0.01\ V/^{\circ}C \cdot 25^{\circ}C\right) + bV$$
$$0.750V - \left(0.01\ V/^{\circ}C \cdot 25^{\circ}C\right) + bV$$
$$b = 0.5V$$

So the equation of our line, based on a temperature (x) and voltage (y), is as follows:

$$yV = \left(0.01\ V/^{\circ}C \cdot x^{\circ}C + 0.5V\right.$$

The only problem with this is that it is a function of the temperature and produces a voltage. The input for calibration will be the voltage output from the sensor. You need to get the temperature, so you must swap the equation around:

$$y = m \cdot x + b$$
$$\frac{y - b}{m} = x$$
$$\frac{yV - 0.5V}{0.01V/^{\circ}C} = x^{\circ}C$$
$$100y \cdot yV - 50V = x^{\circ}C$$

This is rather simple and of course converts into a function we can easily implement in Python:

```
def tmp36_v_to_t(volts):
    return 100 * volts - 50
```

Wiring Up The Project

Now that you understand how to relate the voltage you see to a temperature, you need to wire the sensor to the BeagleBone Black. As discussed previously, wiring is relatively straightforward, but looking at the datasheet, there is one recommended addition: a capacitor between the V_{supply} and ground. A capacitor can be thought of as a reservoir for electrical charge. The purpose of the capacitor, in this case, is just like a water tower you may find in your town. If there is more water going into the tower than is being used, there is room to store the water without increasing the pressure down the system. If more is needed than normal, normally causing a drop in pressure, the water tower stores enough to provide the extra supply. Of course, if the incoming water is *very* unregulated, you could still end up with not enough water or too much water pressure.

Similar situations can occur on our power supplies. They are not always regulated to absolute precision. By putting a capacitor just before the input to the temperature sensor, we can account for tiny fluctuations in the supply voltage to smooth it out. These are often referred to as "smoothing capacitors" and can be found sprinkled throughout digital systems. The capacitor can be seen in Figure 10.3, connecting BBB_VDD_3.3V to the BBB_DGND just before the sensor. The value of a capacitor, known as the capacitance, is measured in Farads (F), and usually very low Farad values (such as the 0.1 µF in this case). The two different types of capacitors are polarized and nonpolarized. In this case, we need a nonpolarized capacitor that will have a code stamped on it of 104. Figure 10.4 shows the breadboard connection diagram.

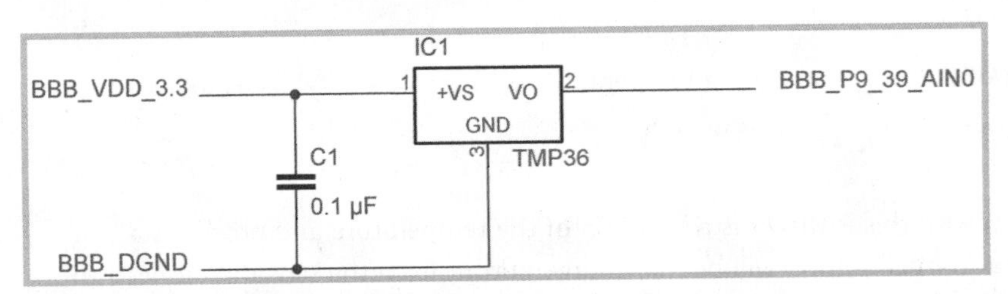

FIGURE 10.3 Schematic for connecting the TMP36 to the BeagleBone Black.

Now you need to collect data with some Python code. The code in Listing 10.1 should be very familiar to you and includes the calibration function just mentioned.

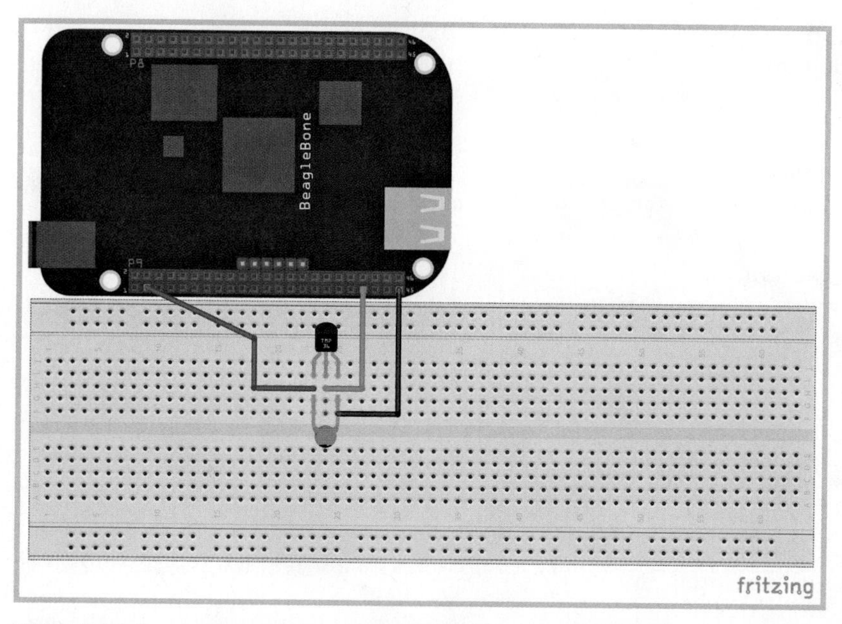

FIGURE 10.4 A TMP36 breadboard diagram for connection to the BeagleBone Black.

LISTING 10.1 tmp36_collection.py

```
import Adafruit_BBIO.ADC as ADC
import time

# Define program constants
TMP36_PIN   = 'AIN0'
SAMPLE_RATE = 0.5      # Hertz

def read_adc_v(adc_pin, adc_max_v=1.8):
    """ Read a BBB ADC pin and return the voltage.

    Keyword arguments:
    adc_pin    -- BBB AIN pin to read (required)
    adc_max_v -- Maximum voltage for BBB ADC (default 1.8)

    Return:
    ADC reading as a voltage

    Note: Read the ADC twice to overcome a bug reported in the
    Adafruit_BBIO library documentation.
```

```
    """
    ADC.read(adc_pin)
    return ADC.read(adc_pin) * adc_max_v

def tmp36_v_to_t(volts):
    """ Calibration function for the TMP36 Temperature sensor.

    Keyword arguments:
    volts - TMP36 reading in Volts

    Return:
    Reading in degC
    """
    return (100 * volts) - 50

if __name__ == '__main__':
    # Configure the ADC
    ADC.setup()

    # Execute until a keyboard interrupt
    try:
        while True:
            voltage_reading = read_adc_v(TMP36_PIN)
            temperature = tmp36_v_to_t(voltage_reading)
            print 'Temperature: {:.2f} C'.format(temperature)
            time.sleep(1/SAMPLE_RATE)

    except KeyboardInterrupt:
        pass
```

This is a little more complex than the code we have worked with in the past, but we are now working toward building a bigger application and need to take into account maintainability, documentation, and good coding practice. First, notice that compared to the Pulse Sensor code in the previous chapter, we have slowed the sample rate on line 6 significantly to 0.5Hz (or a sample every 2 seconds). Temperature is relatively very slow to change, so even this rate can still be considered too fast. We refine sample rates later.

Just after the two function definitions are blocks of comment text between triple quotation marks. This is a special kind of comment called a docstring. A well-written docstring not only makes it clear how a function works in the source code but can also help a developer who may use your code as a library.

The line `if __name__ == '__main__':` tells the Python interpreter to only execute the code that follows if the code is being called directly from the interpreter. This is useful if someone wants to use our code as a library. For example, if you started the Python interpreter in interactive mode, you could enter `import tmp36_collection` just like any other Python library and, with the special `if __name__` call, the functions and constants will be defined but the program will not start executing. This way, you can test out and experiment with the functions if need be or use them elsewhere. This is also where docstrings can be very helpful.

```
root@beaglebone:/bbb-primer/chapter10# python
Python 2.7.3 (default, Mar 14 2014, 17:55:54)
[GCC 4.6.3] on linux2
Type "help", "copyright", "credits" or "license" for more information.
>>> from tmp36_collection import *
>>> read_adc_v(TMP36_PIN)
0.6979999959468842
>>> help(read_adc_v)

Help on function read_adc_v in module tmp36_collection:

read_adc_v(adc_pin, adc_max_v=1.8)
    Read a BBB ADC pin and return the voltage.

    Keyword arguments:
    adc_pin   -- BBB AIN pin to read (required)
    adc_max_v -- Maximum voltage for BBB ADC (default 1.8)

    Return:
    ADC reading as a voltage

    Note: Read the ADC twice to overcome a bug reported in the
    Adafruit_BBIO library documentation.
(END)
```

Finally, there is a new feature on line 44 with the `print` statement. This is a method for string formatting that we won't review in depth here, but you should look up Python string formatting to better understand what is happening in this statement. The address http://docs.python.com is a great place to start looking for information on Python strings. The information between the curly brackets is replaced with the variables in the `format` function call.

Executing the program with the TMP36 sensor wired in as shown in Figures 10.3 and 10.4 results in the following execution results:

```
root@beaglebone:/bbb-primer/chapter10# python tmp36_collection.py
Temperature: 19.80 C
```

```
Temperature: 19.80 C
Temperature: 19.80 C
^Croot@beaglebone:/bbb-primer/chapter10#
```

Seeing the Light

This is a good start. Next, you will add in the light sensor and produce code to provide a similar output. A photocell, shown in Figure 10.5, has a variable resistance based on the light striking the sensor.

FIGURE 10.5 A photocell.

This is similar to the potentiometer. You will take advantage of it to create a voltage divider where the photocell is the first resistor. Figure 10.6 shows the schematic for the photocell added to the schematic in Figure 10.3, and Figure 10.7 shows the photocell added into the breadboard configuration.

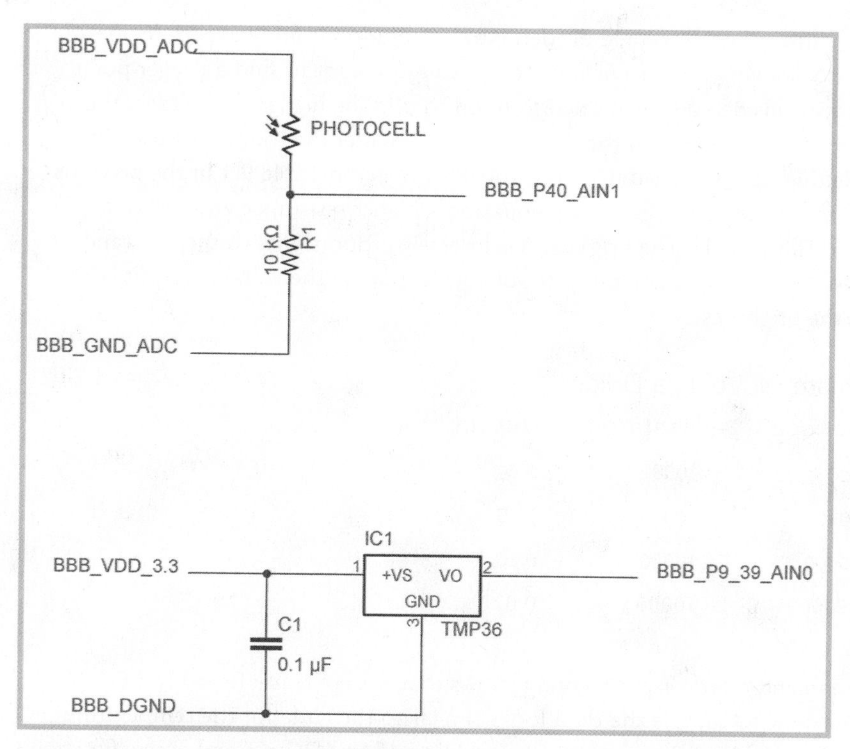

FIGURE 10.6 Photocell added to the schematic for the project.

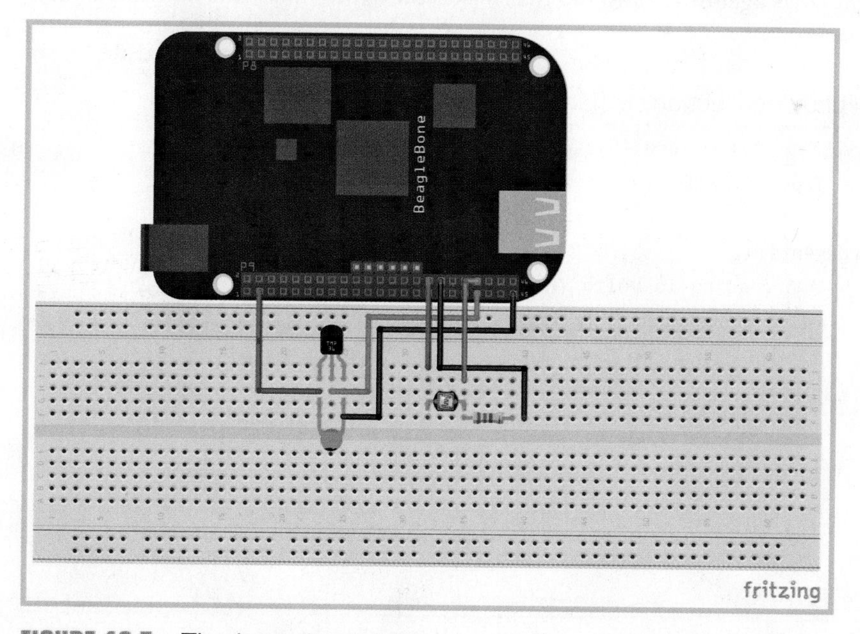

FIGURE 10.7 The breadboard diagram of the project with photocell and temperature sensor.

We use a 10 kΩ resistor for the voltage divider. Now calibration is fairly easy this time because you are not looking for specific light levels; you just want to find a trigger point where the lights are considered transitioned from on to off. The higher the voltage, the brighter the room. As you can see in the datasheet, the brighter the room is, the lower the resistance. Setting up a table similar to the potentiometer in Table 9.1 in the previous chapter, you see that via the voltage divider equation a higher resistance equates to a lower voltage (see Table 10.1). The brightness is inversely proportional to the resistance, the resistance is inversely proportional to the voltage output, so the voltage output is proportional to the brightness.

TABLE 10.1 Photocell Voltage Divider

Input	Photocell	Resistor	Output
1.8	1000	10000	1.64
1.8	10000	10000	0.90
1.8	100000	10000	0.16
1.8	1000000	10000	0.02

In the test environment, a little experimenting shows that the lights are found to be "off" below 0.15V. The code to sample the data looks similar to the code for the temperature sensor in Listing 10.1, with two exceptions: the addition of an ADC pin and a new function to check the light levels against a threshold that makes the lights "on." This function alone is shown in Listing 10.2.

LISTING 10.2 photo_collection.py (Lines 27–36)

```
def lights_on(volts, threshold=0.15):
    ''' Returns True or False is the lights are on or off

    Keyword arguments:
    volts      - ADC reading in volts (Required)
    threshold - value above which the lights are off

    Return:
    Boolean of light status
    '''
    if volts > threshold:
        return True
    else:
        return False
```

Now that you know how to capture both pieces of data, you can begin to see an overall data-collection program coming together. We call this program environment_monitor.py, and the only differences between it and the other programs for collecting data are the additions of the calibration function in Listing 10.1 and the light check function in Listing 10.2 and the collection/output of the data in the main program execution. We should also change the sample rate to once every 5 minutes, or 0.0033Hz. We really don't need to have up-to-the-second information for parameters that don't change that often.

```
temperature = tmp36_v_to_t(read_adc_v(TMP36_PIN))
lights_status = lights_on(read_adc_v(PHOTO_PIN))
print 'Temperature: {:.2f} C      Lights: {}'.format(temperature,
                                                        lights_status)
time.sleep(1/SAMPLE_RATE)
```

Now that you have a small program that sits and reads the sensors every 2 seconds and outputs the data to the console, you need to determine whether you are happy. You'll likely want to do more. According to the check list at the beginning of this chapter, you have two more goals: to publish the data to a storage space on the Internet and to start collecting data.

Publishing the Sensor Data

The next thing to do is to publish the data to the Internet. There are a number of ways to accomplish this goal, and it can get pretty advanced. As a show of the basics, we are going to go with a very simple method.

SparkFun offers a data service at http://data.sparkfun.com. This website provides a very simple interface to push data to the SparkFun servers, where it can be readily accessible for viewing on the Web or to be ingested into other applications.

> **NOTE**
>
> **Security Warning**
>
> The data on this server is publicly accessible on the Internet. You should not put anything on the site that you wouldn't mind telling to a total stranger.

How do you get started publishing your data? First, you need to go to the website and create a stream by clicking the "Create" button, as shown in Figure 10.8.

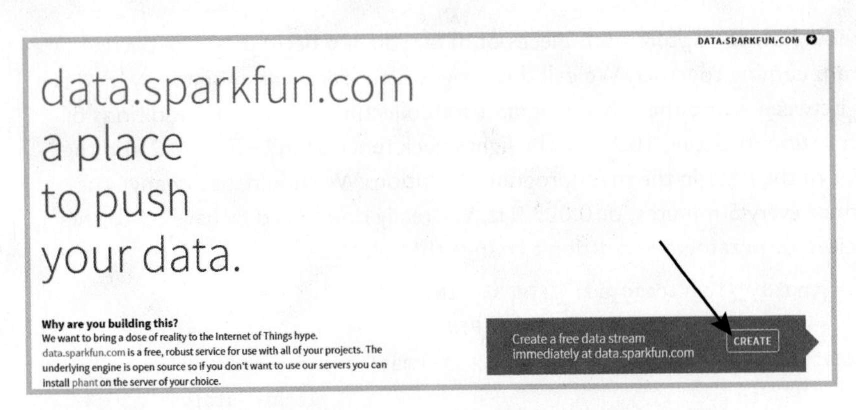

FIGURE 10.8 Create a data stream on data.sparkfun.com.

Clicking the "Create" button will take you to a page where you fill out the following fields (the inputs shown are for the sample build):

- **Title**—BeagleBone Black Primer Environment Monitor
- **Description**—Temperature and light status for the author's workshop
- **Show in Public Stream List?**—Visible
- **Fields**—temperature, lights_on
- **Stream Alias**—Bbb_primer_chapter10
- **Tags**—tutorial, beagleboneblack
- **Location**—Columbia, MD

Once you have entered this information, click Save. The next screen provides a lot of important information. I recommend selecting the option to have the information emailed to you in order to save this information. The fields are as follows: <s$Ipublishing;environment_monitor.py results>

- Public URL
- Public Key
- Private Key
- Delete Key

You'll also see some examples of how these strings are used. The private key and the delete key are important because they are what allow you to update your data stream and delete the data stream, respectively. In addition, you find some examples on these pages for posting data, but you are going to go an even easier route. You'll use a Python library for interacting with the Phant server. Phant is the underlying technology that runs the data.sparkfun.com website. For this to work, you first need to install the python-phant library on your BeagleBone Black using pip on the command line, as shown here:

```
root@beaglebone:/bbb-primer/chapter10# pip install phant
Downloading/unpacking phant
  Downloading phant-0.4.tar.gz
```

```
    Running setup.py egg_info for package phant

  Downloading/unpacking requests (from phant)
    Downloading requests-2.5.1.tar.gz (443Kb): 443Kb downloaded
    Running setup.py egg_info for package requests

  Installing collected packages: phant, requests
    Running setup.py install for phant

    Running setup.py install for requests

  Successfully installed phant requests
  Cleaning up...
```

You can see that this not only installed the phant library but the requests library as well. The requests library provides the HTTP interface that the phant library utilizes to make everything clean and easy to use. The code makes the process so easy that you can go ahead and drop some new code into the environment_monitor.py program and start logging fairly quickly. The python-phant website shows some very straightforward examples (http://github.com/matze/python-phant).

If we look through the code for environment_monitor.py (Listing 10.3) you'll see that first, you need to include the phant library:

```
import phant
```

You then add the public and private keys you previously issued to the source code. Add these to the constants at the top of the source code just to make things easy to find and change later if necessary.

```
PHANT_PRIVATE_KEY = '1zPWybq77pFevXD4gYJV'
PHANT_PUBLIC_KEY  = 'RMGJoAgbbqiGnRd64bLM'
```

Next, we want to add code to our main program execution to initialize the Phant library.

```
sparkfun_data = phant.Phant(PHANT_PUBLIC_KEY,
                    'temperature', 'lights_on',
                    private_key=PHANT_PRIVATE_KEY)
```

This is all you have to change. You now have a working program, shown in Listing 10.3, that collects data and pushes it to the Web! There is also an added conversion from Celsius to Fahrenheit for my own local temperature standard. The conversion can be skipped by simply commenting out the conversion line. As a note for adaptation, there is a rate limit on how often data can be posted to the data.sparkfun.com server: only be 100 data points can be posted every 15 minutes, for an average of just over 9 seconds between logging points. This could also come as a burst of 100 every 15 minutes.

LISTING 10.3 environment_monitor.py

```
import Adafruit_BBIO.ADC as ADC
import time
import phant

# Configure the ADC
ADC.setup()

# Define program constants
TMP36_PIN    = 'AIN0'
PHOTO_PIN    = 'AIN1'

PHANT_PRIVATE_KEY = 'lzPWybq77pFevXD4gYJV'
PHANT_PUBLIC_KEY  = 'RMGJoAgbbqiGnRd64bLM'

SAMPLE_RATE = 0.0033  # Hertz

def read_adc_v(adc_pin, adc_max_v=1.8):
    ''' Read a BBB ADC pin and return the voltage.

    Keyword arguments:
    adc_pin    -- BBB AIN pin to read (required)
    adc_max_v -- Maximum voltage for BBB ADC (default 1.8)

    Return:
    ADC reading as a voltage

    Note: Read the ADC twice to overcome a bug reported in the
    Adafruit_BBIO library documentation.
    '''
    ADC.read(adc_pin)
    return ADC.read(adc_pin) * adc_max_v

def tmp36_v_to_t(volts):
    ''' Calibration function for the TMP36 Temperature sensor.

    Keyword arguments:
    volts - TMP36 reading in Volts

    Return:
    Reading in degC
```

```
    '''
    return (100 * volts) - 50

def lights_on(volts, threshold=0.15):
    ''' Returns True or False is the lights are on or off

    Keyword arguments:
    volts     - ADC reading in volts (Required)
    threshold - value above which the lights are off

    Return:
    Boolean of light status
    '''
    if volts > threshold:
        return True
    else:
        return False

def celsius_to_fahrenheit(degrees_celsius):
    ''' Returns the input celsius temperature as fahrenheit

    Keyword arguments:
    degrees_celsius - ADC reading in volts (Required)

    Return:
    Temperature in fahrenheit
    '''
    return (degrees_celsius * 1.8) + 32

if __name__ == '__main__':

    # Execute until a keyboard interrupt
    try:
        sparkfun_data = phant.Phant(PHANT_PUBLIC_KEY,
                                    'temperature', 'lights_on',
                                    private_key=PHANT_PRIVATE_KEY)

        while True:
            temperature = tmp36_v_to_t(read_adc_v(TMP36_PIN))
            temperature = celsius_to_fahrenheit(temperature)
            lights_status = lights_on(read_adc_v(PHOTO_PIN))
            sparkfun_data.log(temperature, lights_status)
```

```
            time.sleep(1/SAMPLE_RATE)

    except KeyboardInterrupt:

        pass
```

Start Collecting Data

You are now down to one more goal: start collecting data and publishing it automatically when the BeagleBone Black is powered on.

Now you just need the program to execute when the BeagleBone Black is powered on. The first thing you need to do is already shown in Listing 10.3, on a new line that was added to the very top:

```
#!/usr/bin/env python
```

With that added, you can make your python file a standalone executable so that you don't need to invoke the Python interpreter to begin execution. You make it an executable with the `chmod` command, which changes a file's permissions. With the permission set to "executable," Linux is smart enough to look at the new first line and execute it with the Python interpreter automatically. The `755` in the command is an octal number (this is a base-8 number system) that sets the right permission bits to make the file an executable:

```
root@beaglebone:/bbb-primer/chapter10# chmod 755 environment_monitor.py
```

The next step is to tell Linux to start this program as a service when the system boots. You need to write another source file that speaks the language of the Linux service system, copy that file to the `/etc/init.d/` directory, and make it an executable as well (see Listing 10.4). We will not get into the details of writing shell scripts, but I will highlight a couple specific lines of interest toward the end of the process.

LISTING 10.4 environment_monitor.sh

```
#!/bin/sh

### BEGIN INIT INFO
# Provides:           environment_monitor.sh
# Required-Start:     $all
# Required-Stop:      $all
# Default-Start:      2 3 4 5
# Default-Stop:       0 1 6
# Short-Description: Monitor environmental variables and publish them
# Description:        Monitor environmental sensors connected the
#                     BeagleBone Black and push the data to
```

```
#                     data.sparkfun.com
### END INIT INFO

####################################################################
# environment_monitor.sh service start script
#
# Adapted from Stephen C Philips' example script. For starting a
# Python script as a service. (http://blog.scphillips.com/)
#
####################################################################

# Change the next 3 lines to suit where you install your script and
# what you want to call it
DIR=/root/bbb-primer/chapter10/
DAEMON=$DIR/environment_monitor.py
DAEMON_NAME=environment_monitor

# Add any command line options for your daemon here
DAEMON_OPTS=""

# This next line determines what user the script runs as.
# Root generally not recommended but necessary if you are using the
# Adafruit GPIO library from Python.
DAEMON_USER=root

# The process ID of the script when it runs is stored here:
PIDFILE=/var/run/$DAEMON_NAME.pid

. /lib/lsb/init-functions

do_start () {
    log_daemon_msg "Starting system $DAEMON_NAME daemon"
    start-stop-daemon --start --background --pidfile $PIDFILE \
                      --make-pidfile --user $DAEMON_USER \
                      --chuid $DAEMON_USER --startas $DAEMON \
                      -- $DAEMON_OPTS
    log_end_msg $?
}
do_stop () {
    log_daemon_msg "Stopping system $DAEMON_NAME daemon"
    start-stop-daemon --stop --pidfile $PIDFILE --retry 10
    log_end_msg $?
```

```
}

case "$1" in

    start|stop)
        do_${1}
        ;;

    restart|reload|force-reload)
        do_stop
        do_start
        ;;

    status)
        status_of_proc "$DAEMON_NAME" "$DAEMON" && exit 0 || exit $?
        ;;
    *)
        echo \
        "Usage: /etc/init.d/$DAEMON_NAME {start|stop|restart|status}"
        exit 1
        ;;

esac
exit 0
```

The next step is to copy this file into the /etc/init.d directory and make it executable as well.

```
root@beaglebone:/bbb-primer/chapter10# cp environment_monitor.sh \
/etc/init.d
root@beaglebone:/bbb-primer/chapter10# cd /etc/init.d
root@beaglebone:/etc/init.d# chmod 755 environment_monitor.sh
```

The final step to turn our program into a system service is to install it as a service with a program called `insserv`. This program reads in the special comments on lines 3 through 12 of Listing 10.4. These lines provide information to the service system on various properties of program execution. Most importantly, `required-start` and `required-stop`, which have the keyword `$all`, tell the system that everything else that is scheduled should be started first and then our service can be started. We call the `insserv` program first with the `-n` argument to test the installation without actually changing anything and make sure we don't see any errors. If that looks good, we run the same command without the `-n`, as shown here:

```
root@beaglebone:/etc/init.d# insserv -n \
/etc/init.d/environment_monitor.sh
insserv: enable service ../init.d/environment_monitor.sh ->
```

```
/etc/init.d/../rc0.d/K01environment_monitor.sh
insserv: enable service ../init.d/environment_monitor.sh ->
/etc/init.d/../rc1.d/K01environment_monitor.sh
insserv: enable service ../init.d/environment_monitor.sh ->
/etc/init.d/../rc2.d/S06environment_monitor.sh
insserv: enable service ../init.d/environment_monitor.sh ->
/etc/init.d/../rc3.d/S06environment_monitor.sh
insserv: enable service ../init.d/environment_monitor.sh ->
/etc/init.d/../rc4.d/S06environment_monitor.sh
insserv: enable service ../init.d/environment_monitor.sh ->
/etc/init.d/../rc5.d/S06environment_monitor.sh
insserv: enable service ../init.d/environment_monitor.sh ->
/etc/init.d/../rc6.d/K01environment_monitor.sh
insserv: dryrun, not creating .depend.boot, .depend.start,
and .depend.stop
root@beaglebone:/etc/init.d# insserv \
/etc/init.d/environment_monitor.sh
```

As a final step, reboot the BeagleBone Black. When the system is done booting, you can check to see if the service is running by using the status feature of our service script:

```
root@beaglebone:/etc/init.d# shutdown -r now && logout

login as: root
Debian GNU/Linux 7

BeagleBoard.org BeagleBone Debian Image 2014-05-14

Support/FAQ: http://elinux.org/Beagleboard:BeagleBoneBlack_Debian
Last login: Sat Feb 21 16:23:48 2015 from titan.home
root@beaglebone:# /etc/init.d/environment_monitor.sh status
environment_monitor.service - LSB: Monitor environmental variables
and publish them.
          Loaded: loaded (/etc/init.d/environment_monitor.sh)
          Active: active (running) since Fri, 16 May 2014 03:34:35
 +0000; 9 months and 7 days ago
         Process: 581 ExecStart=/etc/init.d/environment_monitor.sh
 start (code=exited, status=0/SUCCESS)
          CGroup: name=systemd:/system/environment_monitor.service
                  └ 684 python /root/bbb-primer/chapter10//environ

ment_mon...
```

The service reports as active and running! Now you should check the data.sparkfun.com website using the URL provided with the URLs and keys given to you during the setup, as shown in Figure 10.9.

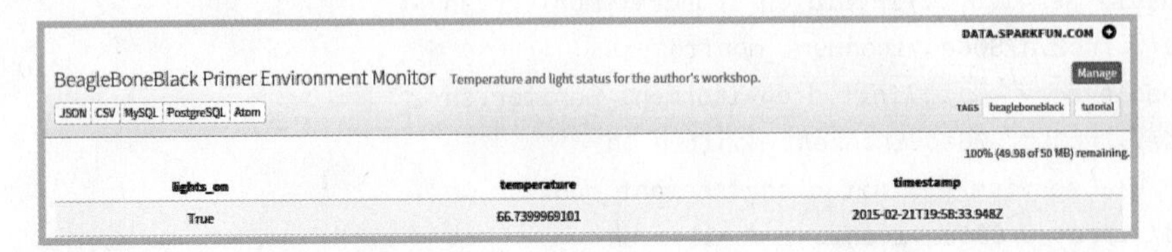

FIGURE 10.9 A line of data from our program sent to the data.sparkfun.com website.

With this, you can even select the CSV button in the upper left and download the entire data collection. The data format makes it easy to import into software, such as Microsoft Excel, and plot the results over a time period, as shown in Figure 10.10.

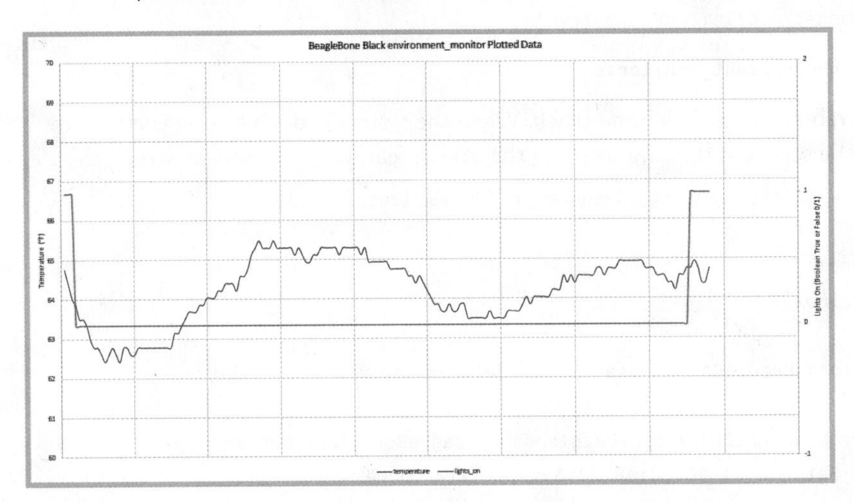

FIGURE 10.10 Data captured from the environment_monitor program plotted out.

If you decide you need to uninstall the service cleanly, you only need to stop the service, remove it with `insserv`, and delete the file from the `/etc/init.d/` directory, as shown here:

```
root@beaglebone:# /etc/init.d/environment_monitor.sh stop
[....] Stopping environment_monitor.sh (via systemctl):
environment_monitor.serv[ o.
root@beaglebone:# insserv -r /etc/init.d/environment_monitor.sh
root@beaglebone:# rm /etc/init.d/environment_monitor.sh
root@beaglebone:# shutdown -r now && logout
```

When the system comes back from the reboot, the data collection will not have started.

Now that you have a firm grasp of how to collect some data about the world around you, you can look to how to manipulate the world around you. In the next chapter, you learn more about driving actuators and indicators.

Interacting with Your World, Part 2: Feedback and Actuators

We've reviewed a lot of information about sensing the world around you and different ways to take data-based actions. Now we will look at how to take action in the physical world with actuators. At the beginning of Chapter 9, "Interacting with Your World, Part 1: Sensors," you learned that a sensor is a type of transducer. For review, we established that a transducer is a device that takes energy as input and then outputs it as another form of energy. With sensors, you take a physical phenomena and transform it into some form of electrical signal. In an actuator, the energy flows in the other direction, taking an electrical signal and converting it to a physical phenomenon.

You've already worked with some basic forms of this transfer with light. You have utilized an LED multiple times and even used a liquid crystal display (LCD) to convey information and status. However, something important related to feedback and actuators, particularly on a board such as the BeagleBone Black, that we haven't covered yet is the transistor. Therefore, we kick off this chapter with a discussion of transistors.

Controlling Current

Remember all the concerns in earlier chapters about the current utilization on the General Purpose Input/Output (GPIO) pins? We were looking at relatively dim LEDs and using the minimum power possible. What if you wanted to turn on a brighter LED that used more power? It is time to look at another fundamental electronic component: the transistor. Why is this important? Because most feedback systems and actuators need more current than we can supply from the BeagleBone Black's GPIO pins.

You've heard the term *transistor* many times but might not understand what one actually does. There are a lot of ways to look at a transistor to understand its operation, but in the end you can simply think of a transistor as a switch. Let's look at an overly complicated example where we want to turn on a brighter LED via a BeagleBone Black GPIO pin but we don't have a transistor. Figure 11.1 shows an odd-looking schematic that contains a finger graphic.

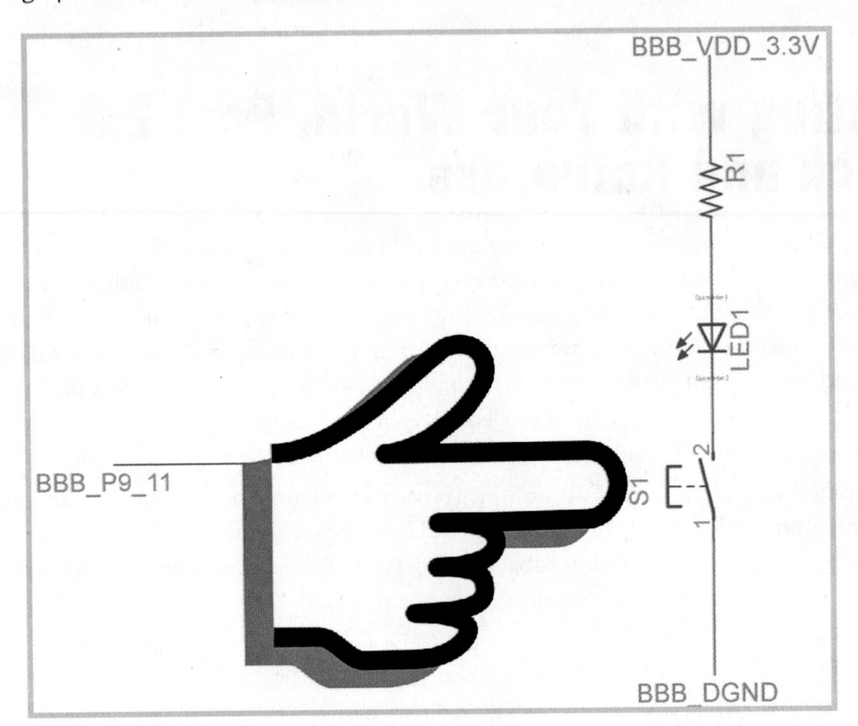

FIGURE 11.1 An odd-looking schematic with a finger.

Obviously a finger graphic is not a standard component for inclusion in a project schematic, but here it is used to trigger a button. Why push the button? Because you can feel a tingle of electricity when P9_11 is turned on. When you feel this tingle as you press and hold the button, electricity flows through the button from the system 3.3V power and turns on the LED. In the much more realistic world in which we live, you can replace the finger graphic and the button with a transistor. Figure 11.2 shows the same idea but using a transistor.

If you follow the metaphor, the idea behind the transistor is easy to understand! When you apply power to one part of the transistor, it enables you to turn on the other part. You just need to learn a couple basics to get going with transistors and work with more power!

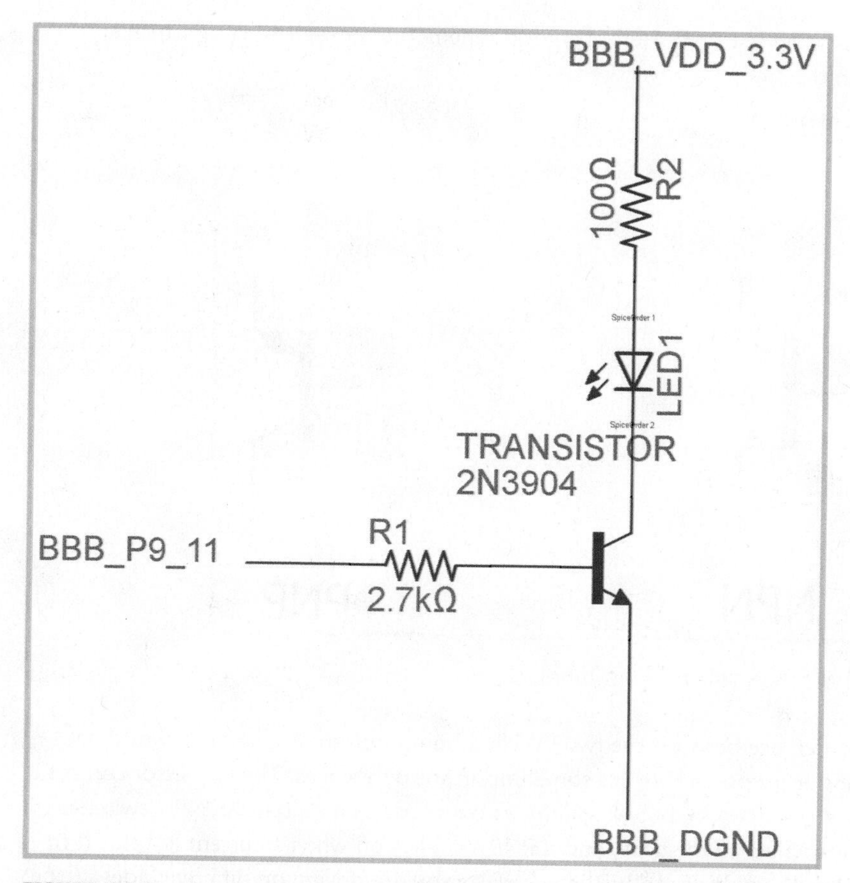

FIGURE 11.2 A schematic that looks much better with a transistor instead of a finger graphic.

As you can see in Figure 11.2, there are three connections to this type of transistor, known as a bi-junction transistor (BJT): the base, the collector, and the emitter. Figure 11.3 displays the two different ways for the three to be shown on a schematic. The base is usually shown as coming in from the left, but not always. Just remember that the base is the part that comes in to the middle of the transistor in the schematic.

In addition, there will always be a connection that has an arrow. This is the emitter. The other part, by process of elimination, must be the collector. The two different types of transistor are differentiated by the swapping of the position of the collector and the emitter. They are referred to by the configuration of semiconductor materials and what material is connected to the base and the collector and emitter. Assuming the base is indicated on the left, the transistor is known as an NPN if the emitter is at the bottom and pointing away from the transistor where N, P, and N refer to the N-type and P-type semiconductor material utilized and how they are used in the three different parts of the transistor.

The other configuration is a PNP. Here is a good mnemonic for remembering which is which is based on the emitter arrow:

```
NPN = Not Pointing iN
```

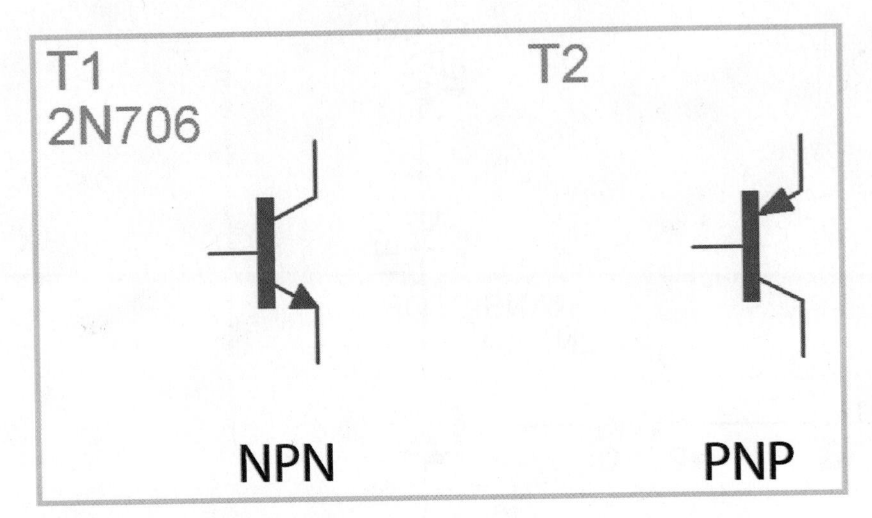

FIGURE 11.3 Two transistor types labeled.

So, what is the difference between the two? Well, a number of things, some of which are in depth, but for our purposes here are some important differences: The transistor can act more like a valve than a switch, but we will focus on its use as a switch. An NPN switches on when a current is applied to the base, and a PNP switches off when a current is applied to the base. Additionally, you want to have the devices you are driving on different sides of the transistor based on whether you are using an NPN or a PNP. With an NPN, the load should come before the transistor, and with a PNP the load should come after the transistor. The load should always be connected to the opposite side of the transistor from the arrow on the symbol. The schematic in Figure 11.4 shows the PNP for comparison against the same type of circuit in Figure 11.2, which uses an NPN.

As discussed previously, you use current to turn the transistor on and off, and you can control the current flow by bringing the GPIO pin high or low, but that just allows current to flow. You need to control how much current flows through by using a resistor, just like you did earlier in the book with the LED. Selecting the right resistor is easy. You need to go back to our $V = I \times R$ equation. Transistors have a turn-on voltage that allows a limited amount of current to flow through the base. Too much current and you not only risk the pin on the BeagleBone Black, you could break the transistor. How do you know the turn-on voltage? By using the datasheet!

A common NPN transistor is the 2N3904, and it needs at least 0.6V applied to turn on, which is called the saturation voltage. However, the transistor can only handle 5mA into the base, and the BeagleBone Black can only handle 8mA. To be safe, you will use just 1mA from the GPIO to control the LED. The actual power for the LED will come from the 3.3V supply pin.

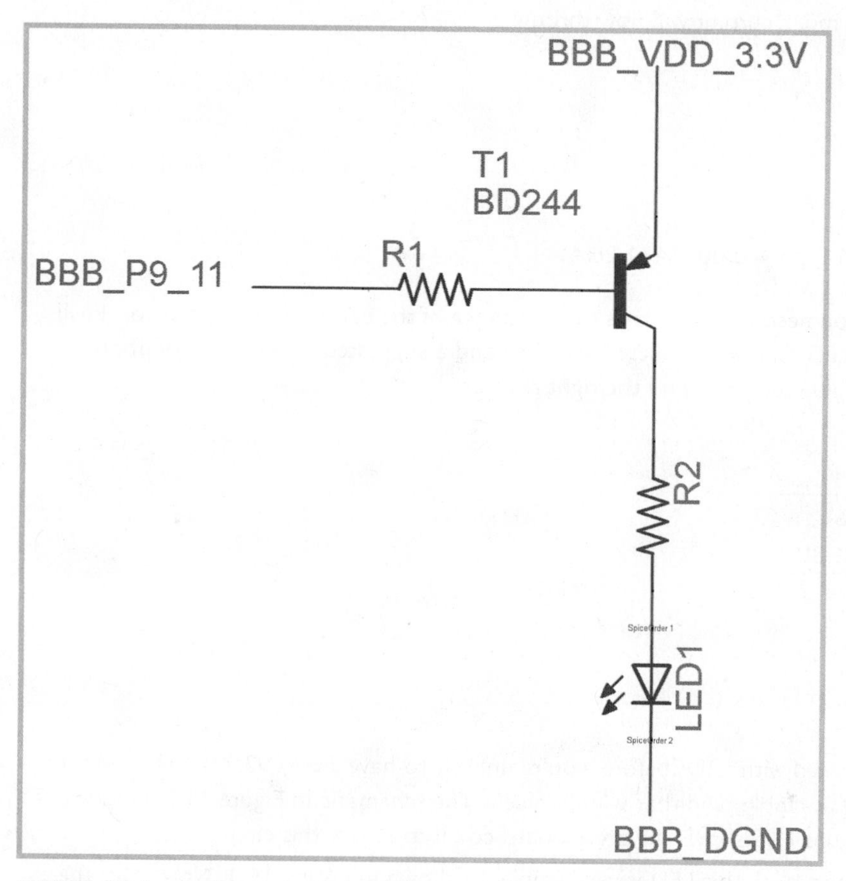

FIGURE 11.4 Using a PNP transistor to drive an LED as opposed to the NPN in Figure 11.2.

You can now apply $V = I \times R$ just like you did for the LED current-limiting resistor in the past:

$$V = I \times R$$

$$R = \frac{V_{source} - V_{saturation}}{I_{max}}$$

$$R = \frac{3.3V - 0.6V}{0.001A}$$

$$R = \frac{2.7V}{0.001A}$$

$$R = 2700\Omega\ (2.7k\Omega)$$

You need a resistor of about 2.7kΩ on the base. At my bench I do have a 2.7kΩ resistor in my kit. Will this do? If you plug the resistance back into the equation from the initial 3.3V, you will see how much current will flow through:

$$V = I \times R$$

$$I = \frac{V}{R}$$

$$R = \frac{3.3V}{2700\Omega}$$

$$I = 0.0012A\ 1.2mI = 0.0012A\ (1.2mA)$$

The next thing you need to calculate is the resistance of the current-limiting resistor. I will use a red 5mm LED that has about a 1.8V drop and a suggested max current of about 16mA. You can plug this in to find the right resistor:

$$V = I \times R$$

$$R = \frac{V_{source} - V_{forward}}{I_{max}}$$

$$R = \frac{3.3V - 1.8V}{0.016A}$$

$$R = \frac{1.5V}{0.016A}$$

$$R = 93.75\Omega$$

As when you worked with LEDs before, you're unlikely to have *exactly* 93.75Ω. However, I do have a 100Ω available, and that will do nicely. The schematic in Figure 11.2 already has these values. Figure 11.5 shows the breadboard connections for this circuit.

The source code to blink this LED is very simple, as shown in Listing 11.1. Note that the frequency on line 5 is set to 1Hz, but the `sleep` statement divides the rate by 2. This is because we consider one period to be the time from the LED coming on to when it is turned on again, or a full cycle.

This may seem trivial compared to some of the work you've accomplished in previous chapters, but the important thing to note is that you are using a bigger and brighter LED than you have previously. A transistor is a powerful companion in our journey. It is at the core of all modern electronics. There are other kinds of transistors than the BJT transistor you are starting with here, such as the MOSFET, which can handle more power and switches on and off faster.

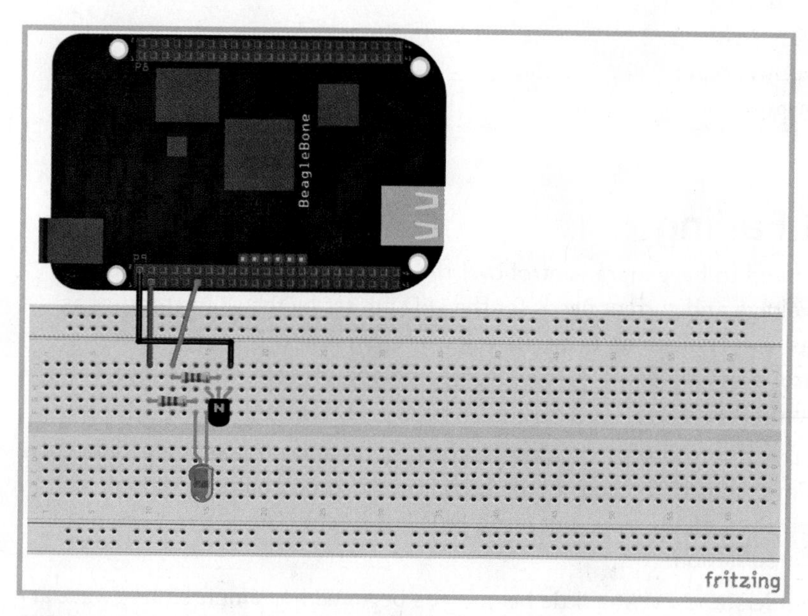

FIGURE 11.5 The breadboard diagram of the circuit in Figure 11.2.

LISTING 11.1 blink.py

```
import Adafruit_BBIO.GPIO as GPIO
import time

LED_PIN   = 'P9_11'
FREQUENCY = 1 # Hz

if __name__ == '__main__':

    state = GPIO.LOW
    GPIO.setup(LED_PIN, GPIO.OUT)
    GPIO.output(LED_PIN, state)

    try:
        while True:
            if state is GPIO.LOW:
                state = GPIO.HIGH
            else:
                state = GPIO.LOW

            GPIO.output(LED_PIN, state)
            time.sleep((1/FREQUENCY)/2)
```

```
except KeyboardInterrupt:
    GPIO.cleanup()
```

Blinking to Fading

Now, what if you wanted to have more control over the LED, such as the ability to fade? You know that the voltage and current needs for the LED are set by the characteristics as reported in the datasheet. You couldn't use less voltage to dim the LED. The answer is to get even better control over how fast the LED is turned on and off with something called pulse-width modulation (PWM).

Imagine, for a moment, that you started switching the LED on and off faster. The difference between on and off may start to become a little less clear. At some point, it will look to your eyes like the LED is on all the time, but just a little less bright. This is the exact idea behind pulse-width modulation.

In order for you to understand PWM a little better, let's begin with a simple square wave at a very specific frequency. If we go for a frequency of 1Hz and connect up the same circuit to a PWM output, we find that we can accomplish what we've tried before with even less code. We'll use P9_14, as shown in Figure 11.6, because it has a PWM output mode. Listing 11.2 shows how simple the code is.

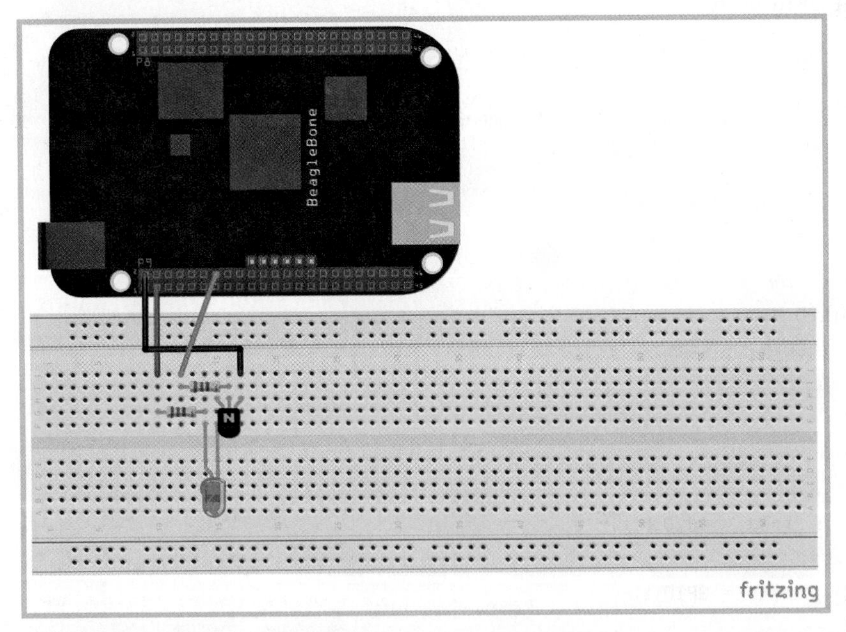

FIGURE 11.6 The breadboard diagram of our blinking circuit with the LED pin moved to P9_14.

LISTING 11.2 pwm_blink.py

```
import Adafruit_BBIO.PWM as PWM

LED_PIN   = 'P9_14'
FREQUENCY = 1  # Hz

if __name__ == '__main__':

    try:
        PWM.start(LED_PIN, 50, FREQUENCY)

        while True:
            pass

    except KeyboardInterrupt:
        PWM.stop(LED_PIN)
        PWM.cleanup()
```

Now that is some simple code! Most of the code is only executed when the program starts up. The `while` loop literally does nothing. The PWM hardware does all the work for us, and you are free to do another activity while the LED happily blinks away. The three parameters to the `PWM.start()` call in Listing 11.2 are the pin desired, the duty cycle, and the frequency. The pins you understand pretty well, and the frequency has already been covered; however, duty cycle is a new term. The duty cycle, as used by the function, defines the percentage of time in a cycle that the pulse is high (in this case, 50%). Figure 11.7 shows an oscilloscope output of the code in Listing 11.2, but with `FREQUENCY` set to 1KHz.

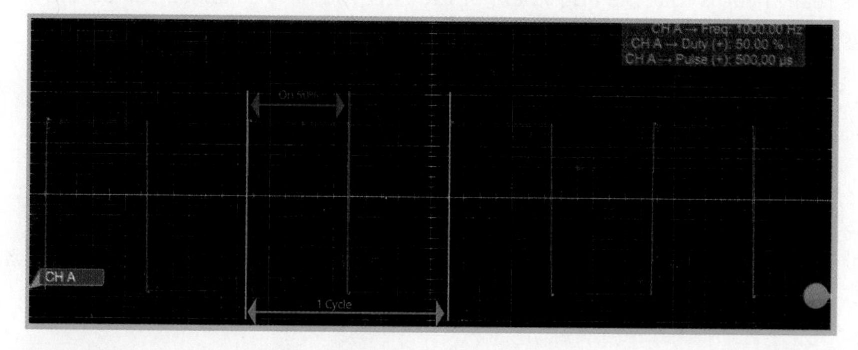

FIGURE 11.7 BeagleBone Black PWM output at 1KHz and 50% duty cycle.

If you set the pulse width to 25%, you will see the pulse high for only 25% of the entire cycle. If you set it to 100%, the pulse would be on all of the time. Figure 11.8 shows the 25% case and a 90% case. The 90% case is shown rather than the 100% case because the 100% case is just a line.

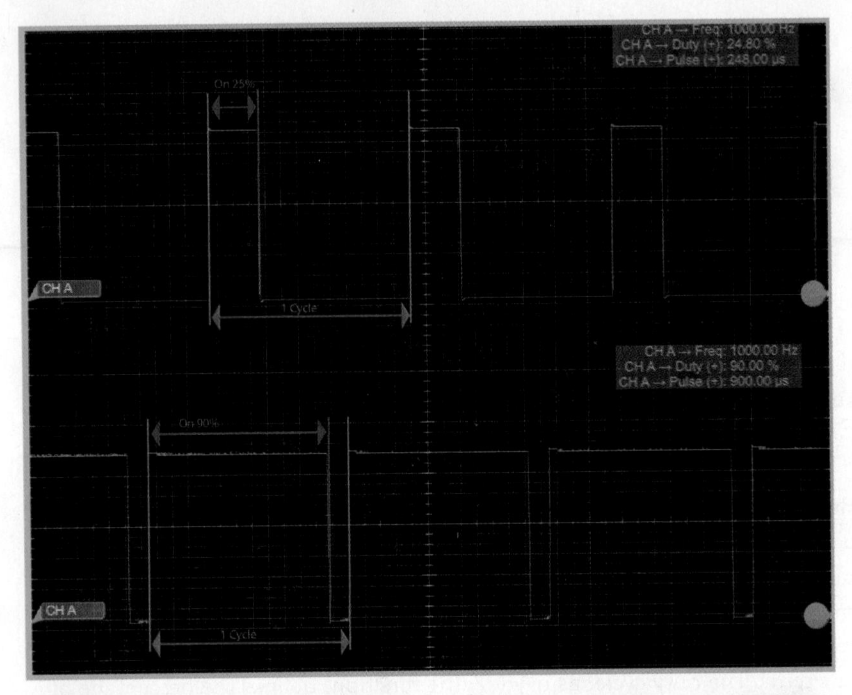

FIGURE 11.8 PWM at 1KHz with a 25% duty cycle and a 90% duty cycle.

What makes this really neat is that the brightness of the LED you have attached will be the same percentage of full-on bright as the duty cycle of the PWM. Thus, you can control the brightness of the LED! In fact, if you combine the code for the blink program with the PWM program, you can fade the LED on and off rather than abruptly switching it on and off (see Listing 11.3).

LISTING 11.3 pwm_fade.py

```
import Adafruit_BBIO.PWM as PWM
import time


LED_PIN = 'P9_14'
PWM_FREQUENCY = 1000  # Hz
BLINK_FREQUENCY = 1   # Hz
```

```
if __name__ == '__main__':
    step_size = 1
    brightness = 0
    fade_wait = ((1/float(BLINK_FREQUENCY))/2) / (100/float(step_size))
    PWM.start(LED_PIN, 0, PWM_FREQUENCY)

    try:
        while True:

            while brightness < 100:
                brightness = brightness + step_size
                PWM.set_duty_cycle(LED_PIN, brightness)
                time.sleep(fade_wait)

            while brightness > 0:
                brightness = brightness - step_size
                PWM.set_duty_cycle(LED_PIN, brightness)
                time.sleep(fade_wait)

    except KeyboardInterrupt:
        PWM.stop(LED_PIN)
        PWM.cleanup()
```

Vibration Motors

Although learning how to use LEDs in a more advanced and safe way is important, the real meat of interacting with the world around us comes from driving a motor and making things move, shimmy, and vibrate. The idea behind motors is simple; you learned all about them in science class. We don't need to go into the physics here, but the basic idea is to run electricity (which you are learning how to manipulate) through a coil of wire and induce a magnetic field. This magnetic field can then work against other forms of magnets and metal to make things move. Simple!

Our first look at a motor will be controlled with a transistor. This is a basic example, and the power for the motor will come from the board supply itself. The motor is a vibration motor like you might find in a cell phone. It is a great way to provide an alert because it can get your attention. The motor is happy with 3.3V and uses only 80mA to 100mA of current. This is too much power for a GPIO pin but is fine for the regular 3.3V power. As you look to use other motors, you will need to supply even more power from off the board.

Figure 11.9 shows the schematic for our vibration motor configuration. There are a couple of interesting items of note here. First, we have added a pull-down resistor to the P9_11 output. Chapter 8, “Low-Level Hardware and Capes,” discussed the fact that the pins actually have a built in, configurable, pull-up and pull-down resistor. The default configuration for P9_11, or GPIO30, is to have this disabled and the receiver off, which means that this pin is floating when you don’t have it exported (in this case, by the Adafruit_GPIO library). This means that the voltage can float just high enough above the ground to allow the transistor to turn on a little bit and leak current through to the motor. Therefore, we add a pull-down for when the pin is not being used by the program. In a more permanent configuration, you could configure the pin with a device tree overlay that is loaded by default and configure the pin to always use a pull-down by default.

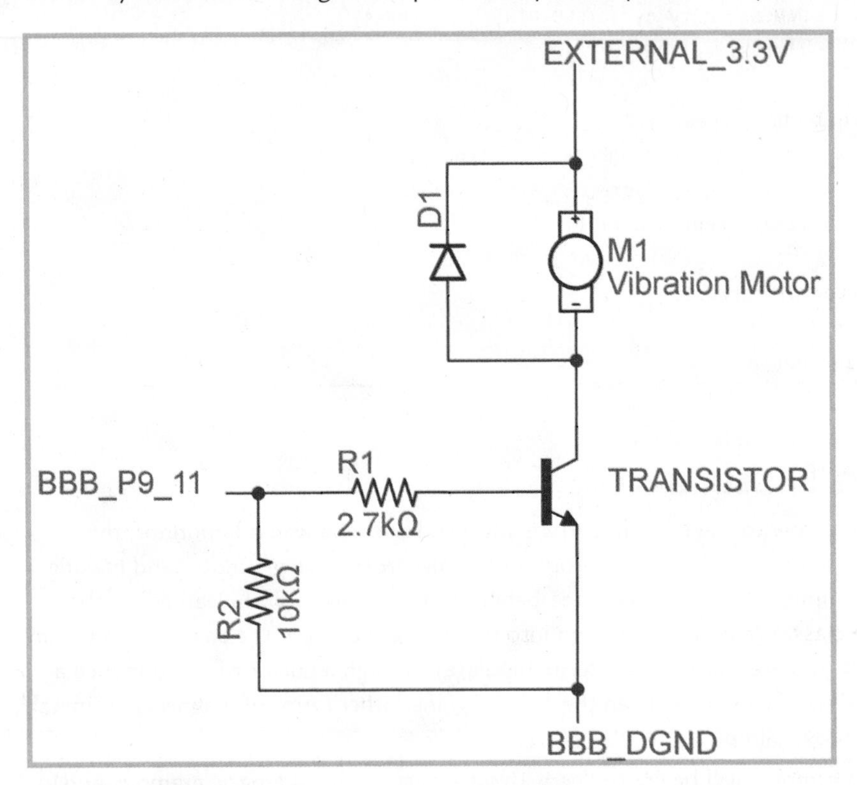

FIGURE 11.9 Basic vibration motor circuit.

Also notice in the Figure 11.9 schematic the item that looks like an LED symbol going back to the input of the motor. This is a general diode that you can think of as a flow control. Current can flow through in only one direction. When you have something like a motor that uses a coil or electromagnet, when the circuit is turned off, the mechanism doesn’t stop instantly. Now that the coil is not driving but our mechanism is moving, the motor becomes a generator with a current becoming induced in the coil. The addition of the diode ensures that the current cycles back through the motor to be dissipated rather than feeding into places you don't want it to go. Listing 11.4 shows our source code for activating this motor.

LISTING 11.4 alert.py

```
import Adafruit_BBIO.GPIO as GPIO
import time

MTR_PIN   = 'P9_11'
FREQUENCY = 1  # Hz

if __name__ == '__main__':

    GPIO.setup(MTR_PIN, GPIO.OUT)
    GPIO.output(MTR_PIN, GPIO.LOW)

    # Pulse for 0.5 seconds
    GPIO.output(MTR_PIN, GPIO.HIGH)
    time.sleep(0.5)
    GPIO.output(MTR_PIN, GPIO.LOW)

    GPIO.output(MTR_PIN, GPIO.LOW)
    GPIO.cleanup()
```

This code is a little different from you've seen before in that it executes once and then exits. This is because we are simply looking to pulse an alert with our motor and not to keep it running. You could copy this into a function defined in another program to simply trigger an alert from within the program. This is about as simple as you can get with a motor of this kind. It is a very small motor with very low current demands. To get into a slightly more complex motor, yet simpler from a circuit perspective, we can use a servo motor.

Servo Motors

A servo motor is useful for many projects because it has a built-in feedback system. This means you can essentially command the motor to a specific position, usually in a range of 180°. It also uses a very simple motor position control system that is fairly close to the pulse-width modulation you used in the LED control previously. In fact, all servos have only three wires: power, ground, and this signal. The system is called pulse-position modulation. The only difference you need to worry about here is that pulse-position modulation is centered on the actual width of the pulse as a function of time rather than a percentage of the pulse width.

Every servo is different, but in general a pulse width of 0.5 milliseconds to 2.5 milliseconds correlates in the servo to 0° to 180° at a pulse frequency of 60Hz. This means that as long as you send a pulse that is 0.5 milliseconds wide every 1/60th of a second, the servo will hold its position at 0°, and likewise for 180° at 2.5 milliseconds. All positions in between are a linear relationship between the two, meaning the math is simple.

The Adafruit_BBIO PWM library that you used to fade an LED before will let you drive a servo that utilizes this kind of control; you just need to translate these pulse durations to the same format the LED fading control utilized, a percentage of a pulse width. This is pretty easy to calculate, with the width of the pulse in seconds:

$$width_{\%} = width_t \cdot frequency$$

This lets us do all the calculations in terms of the pulse durations that are defined in servo datasheets. Note that the pulse time is generally in milliseconds and the equation is in full seconds, so you'll need to convert them. Listing 11.5 shows these calculations and utilizes the PWM library to sweep through a full 180°.

LISTING 11.5 servo.py

```
import Adafruit_BBIO.PWM as PWM
import time

SERVO_PIN = 'P9_14'
ANGLE_RANGE = [0, 180]  # degrees
PPM_RANGE = [0.5, 2.4]  # milliseconds
PPM_FREQUENCY = 50  # Hertz


def ppm_to_pwm(ppm, frequency):
    ''' Convert from a PPM pulse width to a PWM duty cycle

    Keyword arguments:
    ppm        -- PPM pulse width in milliseconds
    frequency -- PPM frequency in Hertz

    Return:
    PWM duty cycle as a percent over 0 to 100
    '''
    return ((ppm / 1000) * frequency) * 100

if __name__ == '__main__':
```

```
# Calculate some variables needed in angle calculation
angle_delta = ANGLE_RANGE[1] - ANGLE_RANGE[0]
ppm_delta = PPM_RANGE[1] - PPM_RANGE[0]
ppm_per_degree = ppm_delta / float(angle_delta)

# Initialize the PWM and go to the lowest PPM
start_position = ppm_to_pwm(PPM_RANGE[0], PPM_FREQUENCY)
PWM.start(SERVO_PIN, start_position, PPM_FREQUENCY)

# Work through the full range of positions
for angle in range(ANGLE_RANGE[0], ANGLE_RANGE[1]):
    relative_angle = angle - ANGLE_RANGE[0]
    relative_ppm = relative_angle * ppm_per_degree
    absolute_ppm = relative_ppm + PPM_RANGE[0]
    duty_cycle = ppm_to_pwm(absolute_ppm, PPM_FREQUENCY)
    PWM.set_duty_cycle(SERVO_PIN, duty_cycle)
    time.sleep(0.1)

PWM.stop(SERVO_PIN)
PWM.cleanup()
```

The ppm_to_pwm function provides the preceding equation conversion with a translation from milliseconds to full seconds and the percentage from a 0-to-1 range to a 0-to-100 range. You can see that it is a 180° servo by the datasheet and the pulse-position modulation values started with the values in the datasheet, but they are tuned to actual numbers by experimentation. Every servo can behave slightly different. Figure 11.10 shows the breadboard implementation with power coming from an external +6V supply. The external supply is used here due to the fact that a servo can quickly start using more power than you want to flow through the BeagleBone Black circuits. Servos may also want voltage levels other than +5V and +3.3 V on the board. The servo used in this example can take up to +6V. The higher the voltage used—within the limits specified in the datasheet—the more force that's applied by the servo. However, as more force is needed, more current will be drawn. This is a standard relationship in motors.

Figure 11.11 shows the circuit as implemented with a bench-top power supply providing the extra power for the servo. In a practical application, such as robotics, power can be provided by other power supplies and even properly sized batteries.

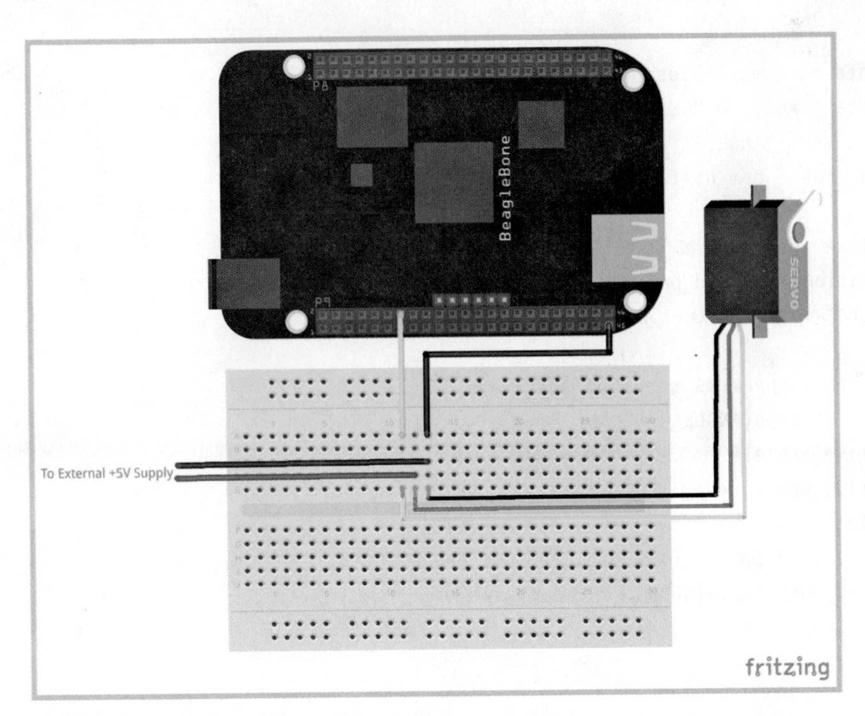

FIGURE 11.10 Servo control breadboard.

FIGURE 11.11 The implemented servo setup. Note the bench-top power supply to provide power to the servo, as well as the smaller servo to show variety.

Stepper Motors

We'll cover one more type of general motor: the stepper motor. The stepper motor can provide more power than many servos—and with extremely precise movements. A stepper motor moves in steps that are defined by the motor coils. In this example, you will use a bipolar stepper motor. A bipolar stepper motor utilizes multiple motor coils, and when a step is executed, the motor stops at that position. The stepper motor used in this example (SparkFun ROB-09238) has the following properties:

- **Step angle**: 1.8°
- **Rated voltage**: 12V
- **Rated current**: 330mA
- **Winding inductance**: 48mH

These are just some of the parameters, but they give us important information for designing our circuit and controlling our motor. A step angle of 1.8° means that when we move a step, the output shaft rotates 1.8°. This means that a full rotation is 200 steps:

$$360°/1.8\ \frac{degrees}{step} = 200360°/1.8\ \frac{degrees}{step} = 200$$

This provides very precise control for the motor position. The control, however, is slightly more complicated. While in the early "getting started" phase of electronics exploration, you can use pre-built drivers. We'll utilize the SparkFun EasyDriver Stepper Motor Driver (SparkFun ROB-13226). This is a great driver because it allows you to simply send a pulse for each step you want to make, and you can configure it for control with 3.3V while accepting a separate power source to drive the motor at the 12V the stepper requires. Figure 11.12 shows a view of the EasyDriver board.

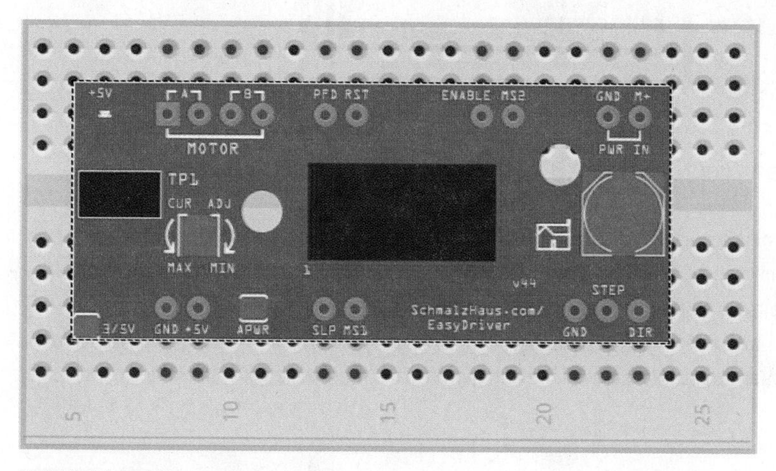

FIGURE 11.12 The EasyDriver.

Take a look at the pins used in the setup. In the top left are four pins that equate to the four leads from a stepper motor, with two pins each for two coils. Some stepper motors have six wires, and you should consult the EasyDriver documentation if your setup has six leads. To the right on the top are the power and ground from the external power source (+12V in our case). In the bottom right are the ground, step signal, and direction from the BeagleBone Black. A logic high or logic low changes direction. The way this is wired shows that logic high is counterclockwise (looking down on the shaft) and logic low is clockwise. Figure 11.13 shows the connections used in this demonstration, and Figure 11.14 shows an actual bench-top setup.

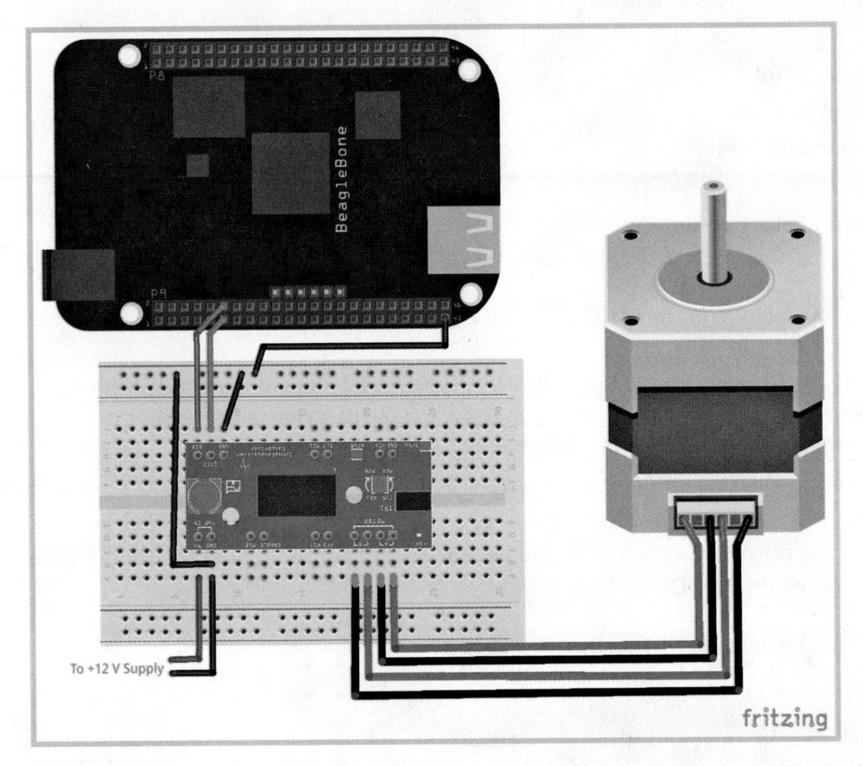

FIGURE 11.13 Stepper motor demonstration breadboard configuration.

Listing 11.6 shows some easy control examples for a stepper motor. Note that although one step is 1.8° on our stepper motor, the EasyDriver allows for microstepping by default, which means there are actually eight step signals per 1.8° increment, or a resolution of 0.225°/pulse. You can actually change this by using more GPIO pins to drive the MS1 and MS2 pins as defined in the datasheet for the control chip (Allegro 3967 Microstepping Driver).

FIGURE 11.14 Stepper motor bench-top implementation.

To ensure you don't overdrive the stepper, you need to look at the parameters again. We already noted that you are going to need a 12V DC external power source, and you know that it needs to be able to supply 330mA. The winding inductance helps you understand how quickly the coils charge up and discharge. For a single step, a coil needs to execute both acts. The minimum amount of time per step is found as follows:

$$time_{step} = \frac{inductance \times current \times 2}{voltage}$$

$$time_{step} = \frac{0.048H \times 0.330A \times 2}{12V}$$

$$time_{step} = 2.64ms$$

This means a single, full step can go no faster than 2.64ms. Also note that if you look at the board in Figure 11.12, you'll see a small potentiometer to limit the current to the motor. The MAX label on the potentiometer refers to a maximum current limiting setting and not a setting that allows the maximum current through the circuit. This has to be dialed back to allow the full 330mA of current through to the stepper.

LISTING 11.6 stepper.py

```
import Adafruit_BBIO.GPIO as GPIO
import time
import ctypes

libc = ctypes.CDLL('libc.so.6')

STEP_PIN = 'P9_11'
DIR_PIN = 'P9_12'
MICROSTEPS = 8


def move(steps, speed):
    ''' Move the stepper so many steps at a specific rate

    Keyword arguments:
    steps -- Number of steps to move
    speed -- Time for a full step (milliseconds)

    Note: This function commands step signal pulses. This
          may not control full step motions depending
          upon any microstepping defined in the control.
    '''
    if steps > 0:
        GPIO.output(DIR_PIN, GPIO.HIGH)
    else:
        GPIO.output(DIR_PIN, GPIO.LOW)

    steps = abs(steps)

    delay = ((float(speed) / 2) / MICROSTEPS) * 1000

    steps_remaining = steps
    while steps_remaining > 0:

        GPIO.output(STEP_PIN, GPIO.HIGH)
        libc.usleep(int(delay))

        GPIO.output(STEP_PIN, GPIO.LOW)
        libc.usleep(int(delay))
```

```
        steps_remaining = steps_remaining - 1


def move_degrees(degrees, speed, degrees_per_step=1.8,
                 usteps_per_step=8):
    ''' Move the stepper so many degrees at a specific rate

    Keyword arguments:
    degrees           -- Number of degrees to move
    speed             -- Time for a full step (milliseconds)
    degrees_per_step -- Number of degrees in 1 full step (1.8 default)
    usteps_per_step  -- Number of microsteps per step in the driver

    '''
    steps = degrees / (degrees_per_step / usteps_per_step)
    move(steps, speed)

if __name__ == '__main__':

    GPIO.setup(STEP_PIN, GPIO.OUT)
    GPIO.output(STEP_PIN, GPIO.LOW)

    GPIO.setup(DIR_PIN, GPIO.OUT)
    GPIO.output(DIR_PIN, GPIO.HIGH)

    try:

        move_degrees(90, 100)
        time.sleep(1)
        move_degrees(-90, 100)
        time.sleep(1)
        move_degrees(360, 10)
        time.sleep(1)
        move_degrees(-360, 2.64)
        time.sleep(1)
        move_degrees(360*2, 3)
        time.sleep(1)
        move_degrees(-360, 2.64)
```

```
except KeyboardInterrupt:
    pass

finally:
    GPIO.output(STEP_PIN, GPIO.LOW)
    GPIO.cleanup()
```

Note that this code utilizes a special call to external libraries to allow for micro-second timing to properly control the speed.

In the servo example and the stepper example, note that the diode from the normal motor example isn't included. The control mechanisms in these examples account for the circuit protection that had to be included.

You have looked at some examples of visual output using light, mechanics, and sound with the vibration motor, and mechanical control with the servo and stepper motors. You can combine different versions and different numbers of these systems for control of something much bigger, such as a robot or even a flying drone. This is the first step toward interacting with the larger world.

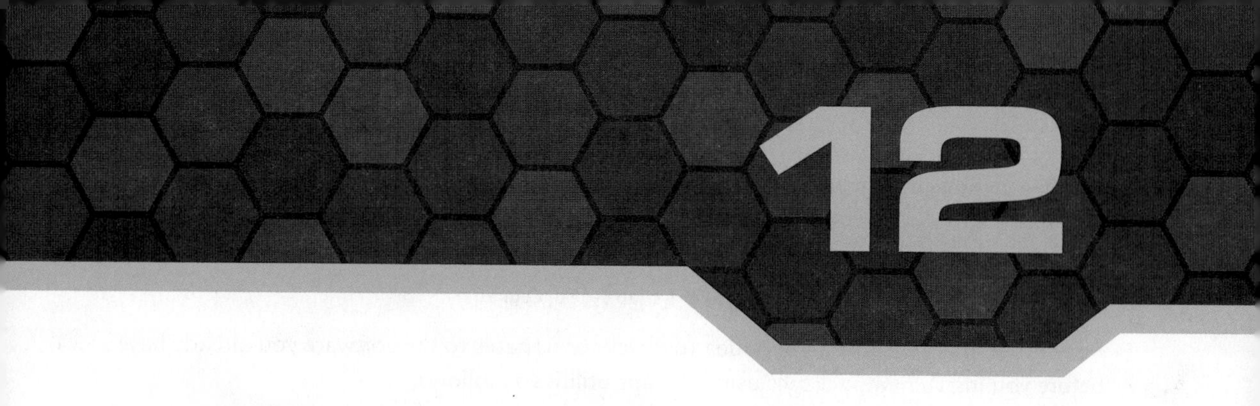

Computer Vision

Many of the projects we've looked at so far show the potential of the BeagleBone Black with hardware interfaces, but now we will look at a project that takes advantage of the fact that the board is a fairly powerful computer in a small package. We will give the BeagleBone Black the gift of sight.

No, we aren't going to discuss the skills necessary to create Vision, the Marvel android superhero—although, that would be fun. In this case, we are discussing the concept of computer vision as the ability for a machine to not only capture an image, or stream of images, but to perform some intelligent operations on them. We need, of course, to start at the beginning and provide our BeagleBone Black a camera.

Connecting a Camera

Webcams that work with Linux—and, in particular, embedded systems such as the BeagleBone Black—are known for being a little hit and miss. However, I've noticed this situation improving over time. In this chapter we will use an off-the-shelf Logitech webcam. There is nothing special in the selection of this webcam other than the fact that it is handy on the workshop shelf. It plugs in to the USB host connection on the far end of the P9 connectors from the power plug. You can see this with the camera plugged in via USB in the foreground of Figure 12.1.

Remember, everything in the Linux environment is in the file system. A video camera will appear in the /dev/ directory as a video device. Without the camera installed, you would not see video devices on the BeagleBone Black. With no camera plugged in, you should see the following:

```
root@beaglebone:~# ls /dev/video*
ls: cannot access /dev/video*: No such file or directory
```

This is expected because there are no video devices built in to the BeagleBone Black. Now, if you plug in a compatible USB camera, you should see a change in the results:

```
root@beaglebone:~# ls /dev/video*
/dev/video0
```

This is a great sign that the camera has successfully connected and is working. You can now move into testing the camera. You install and use a simple command-line utility called streamer to test out the camera:

```
root@beaglebone:~# apt-get update && apt-get upgrade
```

Remember, it is generally a good idea to check for updates to the software you already have before you install new packages using the apt utilities as follows:

```
root@beaglebone:~# apt-get install streamer
Reading package lists... Done
Building dependency tree
Reading state information... Done
The following extra packages will be installed:
  libexplain30 lsof xawtv-plugins
Suggested packages:
  xawtv
The following NEW packages will be installed:
  libexplain30 lsof streamer xawtv-plugins
0 upgraded, 4 newly installed, 0 to remove and 0 not upgraded.
Need to get 811 kB of archives.
After this operation, 1674 kB of additional disk space will be used.
Do you want to continue [Y/n]?
```

You can press Return here and continue with the installation. As you can see, installing streamer requires a number of new libraries and packages to be installed, but the apt tools take care of that for you. With streamer now installed and the camera connected, you can take a test image. Normally you can look to the manual page with `man` to see how to use a utility such as this, but the man page is basic and redirects to much more comprehensive help accessed by running the program with the `-h` option, like so:

```
root@beaglebone:~# streamer -h
```

We won't review the full output here because it is fairly long, but as we go through a couple of examples, you can look at the options in use to understand them better. Looking at the output shows how versatile the stream utility can be. We start with a simple image capture, a BeagleBone Black self-portrait:

```
root@beaglebone:~# streamer -o picture.jpeg
```

The captured image, shown in Figure 12.1, is saved in the execution directory, in this case `/root`.

This is a fairly simple step, but it ensures that a basic structure is in place for the more advanced work to come. You can create a fun application with this basic setup and the skills you already have. How about a simple Python script you can start that connects the BeagleBone Black to a button and takes a picture every time the button is pressed? The breadboard implementation is simple, as shown in Figure 12.2.

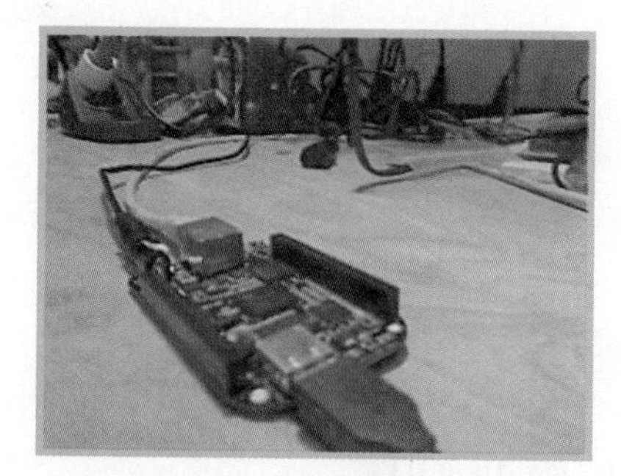

FIGURE 12.1 A BeagleBone Black self-portrait.

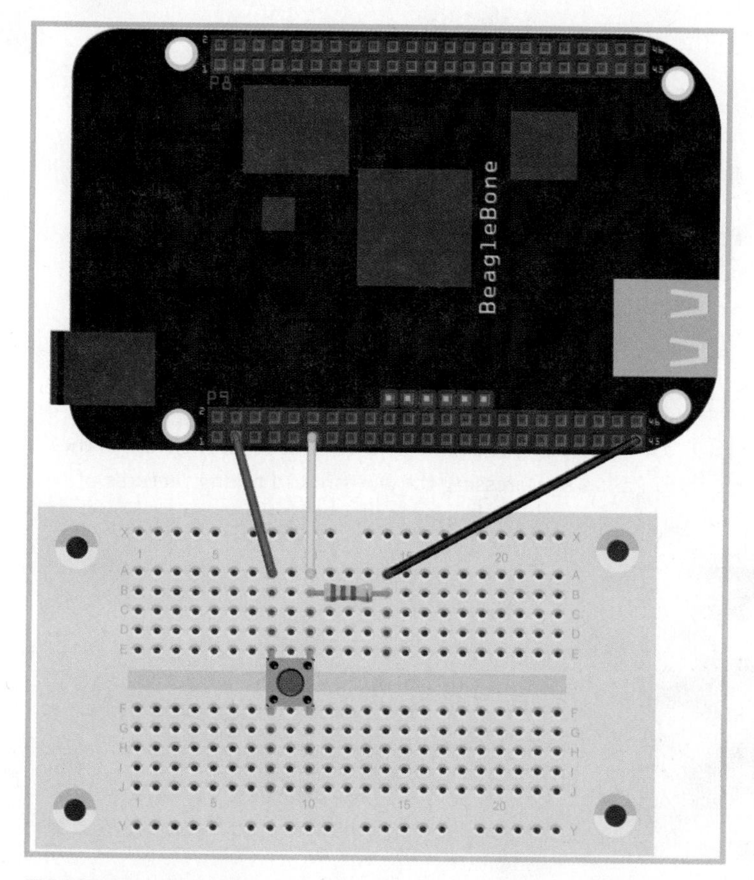

FIGURE 12.2 Button press breadboard example.

The Python code to run the program is simple and uses what you learned in Chapter 9, "Interacting with Your World, Part 1: Sensors," to wait for a button press (see Listing 12.1).

LISTING 12.1 snapshot.py

```
import Adafruit_BBIO.GPIO as GPIO
import subprocess
# Define program constants
BUTTON_PIN = 'P9_11'
# Configure the GPIO pin
GPIO.setup(BUTTON_PIN, GPIO.IN)
if __name__ == '__main__':
    # print out a nice message to let the user know how to quit.
    print('Starting snapshot program, press <control>-c to quit.\n')
    picture_count = 0
    # Execute until a keyboard interrupt
    try:
        while True:
            GPIO.wait_for_edge(BUTTON_PIN, GPIO.RISING)
            output_file = 'snaps/snap{:0>3}.jpeg'.format(picture_count)
            command_call = ['streamer', '-q', '-o', output_file]
            print('Click! Image saved to {}.'.format(output_file))
            subprocess.call(command_call)
            picture_count = picture_count + 1
    except KeyboardInterrupt:
        GPIO.cleanup()
```

This code provides a great basis for more advanced small projects. A good measure is the program's success is how much fun a kid has pressing the button and taking pictures of himself or herself, as in Figure 12.3!

FIGURE 12.3 A couple of kids playing with our snapshot program.

Also, don't forget that this is a video camera. Video capture is as easy as picture capture, as shown next:

```
root@beaglebone:~# streamer -q -f rgb24 -r 60 -t 00:00:10 -o video.avi
```

This is useful, but you are going to want to display to preview the image. Of course, you can accomplish this in two ways. You can hook up a monitor via HDMI, as you did with the BeagleSNES demonstration in Chapter 6, "Trying Other Operating Systems." You can also use a video-based Cape that provides a display. For the examples in this chapter, we will assume the display is showing the standard X-Windows session.

For starting these examples, you will need a keyboard with only one USB host port being used by the camera. You have two options. You can connect a USB hub and then connect a keyboard, the camera, and a mouse, or you can continue to remote control via an SSH connection. The ultimate goal is to have a simple display and interface to use the remote connection option, but remember that a local keyboard is always possible as well.

Now, keep in mind that the goal of this book is to provide a basic introduction to the use of the BeagleBone Black, so we'll look for something to do the hard work for us. To that end, for our enhanced photo booth application, you are going to install a set of libraries that will do the heavy lifting. With the capabilities of these libraries, heavy lifting may be an understatement and the start of Iron Man's JARVIS may be a more apt metaphor. We are going to use OpenCV.

OpenCV stands for Open Computer Vision. To understand what this means, you need to look briefly at what Computer Vision means. At the most basic level, Computer Vision is the capture of image data as well as the ability to perform some set of actions in a seemingly intelligent manner. It overlaps with the field of Artificial Intelligence, but to say that it is a field based just in Artificial Intelligence alone would ignore the diversity of the field. Computer Vision is, in reality, an umbrella term that captures a wide range of fields. Often, the goal is to extract data from a captured image that a machine can interpret in some way so as to provide information about the world around it.

If we think about happens with biological vision, there is far more at work than the simple capture and storage of an image to our brain. We are in a constant state of data processing from all of our senses. As we look around, the image data captured by our eyes undergoes constant analysis. We look for patterns, processing the difference between our eyes to provide depth perception. We process colors based on the combinations of cones that are stimulated in our retina, and we detect changes that indicate movement. This is just a small list of functions, processed as a stream of data, that our biological vision provides.

OpenCV is an open-source toolkit that seeks to provide and refine many of these same functions within a computer system. The pure magnitude of what is provided as a free toolkit in OpenCV is staggering. It is a true gift to the world to have these libraries available. We are going to start with a basic interaction with the OpenCV system and then take a look at some of the more advanced features. Ultimately, we'll teach our BeagleBone Black to keep its eye on us.

Before we can do anything with OpenCV, we must first install the libraries. This isn't a difficult process, but it is more complex than running a simple `apt-get` command. First, we need to make sure we have all the prerequisite libraries installed via `apt-get`. Some of these libraries may already exist on your system, but it is better to let the apt system decide whether or not you have them than to get confused later because you missed a library and are receiving cryptic errors. We will begin by, again, making sure that everything is up to date:

```
root@beaglebone:~# apt-get update && apt-get upgrade
```

This is also a good time to check for a more advanced version of your system upgrades and then perform a cleanup of any libraries that may be going unused. These actions may produce no results, but they are good housekeeping tasks:

```
root@beaglebone:~# apt-get dist-upgrade

root@beaglebone:~# apt-get autoremove
```

Now you need to install some libraries that are necessary. First, you'll install some build tools because you won't just install OpenCV—you will need to build the system for yourself:

```
root@beaglebone:~# apt-get install build-essential cmake git pkg-config
```

As we have already been doing, we will use Python for our OpenCV development, so you will want to make sure you have some additional Python tools available:

```
root@beaglebone:~# apt-get install python-dev python-numpy
```

Next, you need to install some libraries for the graphical user interface (GUI) interactions and some essential media libraries:

```
root@beaglebone:~# apt-get install libgtk2.0-dev libavcodec-dev \
libavformat-dev libswscale-dev libjasper-dev
```

Finally, you need to install some basic image libraries:

```
root@beaglebone:~# apt-get install libjpeg-dev libpng-dev libtiff-dev
```

Once all the libraries are installed, you will need to download the OpenCV system from the repository. The repository is a type of storage known as git, which allows for keeping track of versions of software and for easier collaboration, provides multiple software development branches, and offers many other wonderful features. You can clone it to a local directory with the git command:

```
root@beaglebone:~#  git clone https://github.com/Itseez/opencv.git
```

This will take a few minutes to copy because it is a very large set of files. After that is complete, it is time to actually build and then install the OpenCV system. There will be output from these commands that are not replicated here because they would take pages and pages to show.

```
root@beaglebone:~#  cd opencv
root@beaglebone:~/opencv#  mkdir build
root@beaglebone:~/opencv#  cd build
```

The cmake command gets everything ready to build, setting flags for our BeagleBone Black system so that it is built correctly for your environment:

```
root@beaglebone:~/opencv/build#  cmake -D CMAKE_BUILD_TYPE=RELEASE -D CMAKE_
INSTALL_PREFIX=/usr/local -D WITH_CUDA=OFF -D WITH_CUFFT=OFF -D
WITH_CUBLAS=OFF -D WITH_NVCUVID=OFF -D WITH_OPENCL=OFF -D
WITH_OPENCLAMDFFT=OFF -D WITH_OPENCLAMDBLAS=OFF -D
BUILD_opencv_apps=OFF -D BUILD_DOCS=OFF -D BUILD_PERF_TESTS=OFF -D
BUILD_TESTS=OFF -D ENABLE_NEON=on ..
```

Now you will build the actual library. This next command takes on the order of 90 minutes to complete—a good time to step away, get a cup of coffee or tea, interact with loved ones, or feed your fish:

```
root@beaglebone:~/opencv/build#  make
```

After this has completed, you have now successfully built the OpenCV system. You now need to install it as part of the overall system:

```
root@beaglebone:~/opencv/build#  make install
root@beaglebone:~/opencv/build#  ldconfig
```

Utilizing OpenCV Libraries

You should now have the OpenCV system installed and ready to go. Next, we will write and install a basic video camera program that uses OpenCV to capture data from the camera and display it (see Listing 12.2).

LISTING 12.2 video.py

```
import cv2
if __name__ == '__main__':
    # Define the name of a main window and create it
    main_window_name = 'BeagleBone Black Video'
    cv2.namedWindow(main_window_name, cv2.WINDOW_NORMAL)
    # Configure the video device for capture, -1 indicates
    # the default, in our case, /dev/video0
    video_capture = cv2.VideoCapture(-1)
    try:
        while True:
            # Capture a frame from the camera
            ret, frame = video_capture.read()
            # Display the frame in our window
            cv2.imshow(main_window_name, frame)
            cv2.waitKey(1)
    except KeyboardInterrupt:
        # Clean everything up
        video_capture.release()
        cv2.destroyAllWindows()
```

This is simple code, and the comments spell out everything that is happening. The library import of cv2 is the OpenCV library itself. We then create a window and connect to the video camera. After that, the program simply cycles through reading a frame and displays the frame in the window created. Once you press Ctrl+C, you can do some cleanup.

When you connected the display and started the BeagleBone Black, the display should have booted directly into an X-Windows environment. The user for this environment is debian and the default debian password is temppwd. To have the normal remote terminal execute in the X-Windows environment, you need to switch to the debian user, enable remote display control, and actually make the remote display the active display for the terminal:

```
root@beaglebone:~/bbb-primer/chapter12# su debian
debian@beaglebone:/root/bbb-primer/chapter12$ xhost +
access control disabled, clients can connect from any host
debian@beaglebone:/root/bbb-primer/chapter12$ export DISPLAY=:0;
```

A Better Photo Booth

The code in Listing 12.2 demonstrates a great deal of functionality. What you can do now is improve the photo booth that was started in Listing 12.1. You have a live image preview now to go with the photo booth and a great set of image libraries, so you can integrate these together. Listing 12.3 shows this improved photo booth with a snapshot button! The button is still wired in the same way as our previous circuit, as shown in Figure 12.2.

LISTING 12.3 photobooth.py

```
import Adafruit_BBIO.GPIO as GPIO
import cv2
import time
from datetime import datetime
# Define program constants
BUTTON_PIN = 'P9_11'
MAIN_WINDOW = 'BeagleBone Black Photobooth'
TIMESTAMP_FORMAT = '%Y-%m-%d_%H-%M-%S'
if __name__ == '__main__':
    try:
        # Configure the GPIO pin and add an event watch to check if
        # the button has been pressed
        GPIO.setup(BUTTON_PIN, GPIO.IN)
        GPIO.add_event_detect(BUTTON_PIN, GPIO.RISING)
        # Configure the video device for capture, -1 indicates
        # the default, in our case, /dev/video0
        video_capture = cv2.VideoCapture(-1)
        while True:
```

```
            # Capture frame-by-frame
            ret, frame = video_capture.read()
            # Display the resulting frame
            cv2.imshow(MAIN_WINDOW, frame)
            # If the button has been pressed
            if GPIO.event_detected(BUTTON_PIN):
                # Grab the current time for a timestamp in the filename
                # and generate the file path & name
                snap_time = datetime.now().strftime(TIMESTAMP_FORMAT)
                output_file = 'snaps/snap_{}.jpeg'.format(snap_time)
                # Configure the output to be a JPEG with a 90% image
                # quality
                jpeg_settings = [int(cv2.IMWRITE_JPEG_QUALITY), 90]
                # Write the frame to the file
                cv2.imwrite(output_file, frame, jpeg_settings)
                # Wait for an extra second with the captured frame
                # displayed
                # to give a little feedback to the user
                time.sleep(1)
            cv2.waitKey(1)
except KeyboardInterrupt:
    # Clean-up
    video_capture.release()
    cv2.destroyAllWindows()
```

As you can see, the two programs are combined with a few minor modifications. The main change from the initial snapshot program is the use of the Python `datetime` libraries to generate a time- and date-based label for our snapshot files. This simplifies having to keep track of how many frames you've captured and also pauses the output for a second when the picture is snapped as a little bit of feedback to the user that something has happened.

The BeagleBone Black is truly starting to simulate the kind of expectations you would have for something that has "vision." The OpenCV libraries are expansive, and after you capture an image frame, you can do a number of things. For example, you can take a frame and change the color, find all of the sharp corners, identify various features, and use many more features and transforms. In fact, you can use these features to give BeagleBone Black the capability to identify a face.

Identifying a face will push some of the limits of the computing power of the BeagleBone Black. Running a simple face-recognition program will peg the processing power of the board at the top end and noticeably raise the temperature of the board. For these reasons, to get the best performance you can with a continuous live video feed, you will make a shift toward optimizing the program for performance rather than readability.

Cascade Classifiers

One method used in Computer Vision to identify something in an image is to use a cascade classifier. The idea behind a cascade classifier is simple: Look for a specific set of rough features in the image such as eyes, nose, and mouth, and if there are areas that have those features, additional searches focus on just those areas. This search of the image frame cascades down until there is a reasonable probability that the area identified has a face. This is done at many different scales to look for the feature in question at different sizes within the image. OpenCV comes with a series of classifiers that search for different features with different levels of accuracy. Experimenting with the BeagleBone Black, I've found that the best choice for our platform for basic face recognition is the Local Binary Pattern (LBP) based classifiers. Listing 12.4 has the code for face recognition added to the basic video program.

LISTING 12.4 face_tracker.py

```
import cv2
if __name__ == '__main__':
    # Define the name of a main window and create it
    main_window_name = 'BeagleBone Black Video'
    cv2.namedWindow(main_window_name, cv2.WINDOW_NORMAL)
    # Set up the identification cascade to use
    cascade = '/root/opencv/data/lbpcascades/lbpcascade_frontalface.xml'
    face_cascade = cv2.CascadeClassifier(cascade)
    # Configure the video device for capture, -1 indicates
    # the default, in our case, /dev/video0
    camera = cv2.VideoCapture(-1)
    try:
        while True:
            # Capture frame-by-frame
            ret, frame = camera.read()
            # Scale the frame down, convert it to grayscale,
            # and then search for faces.
            reduced = cv2.resize(frame, (0, 0), fx=0.5, fy=0.5)
            reduced_grey = cv2.cvtColor(reduced, cv2.COLOR_BGR2GRAY)
            faces = face_cascade.detectMultiScale(reduced_grey, 1.3, 2)
            # Iterate over all the faces identified in the image and draw
            # rectangles around them.
            for (x, y, w, h) in faces:
                origin = (x * 2, y * 2)
                size = (w * 2, h * 2)
                far = (origin[0] + size[0], origin[1] + size[1])
                cv2.rectangle(frame, origin, far, (0, 0, 255), 2)
```

```
            # Display the resulting frame
            cv2.imshow(main_window_name, frame)
            cv2.waitKey(1)
    except KeyboardInterrupt:
        # Clean everything up
        camera.release()
        cv2.destroyAllWindows()
```

In Listing 12.4, an XML file is identified that holds the classifier information. This file is used to create a CascadeClassifier object. This object does all the work of face tracking. After the frame is captured, you can convert the image to grayscale because this is the form the classifier needs. The image is also reduced. You do this because the larger the image, the more time the classifier will take. Reducing the image in half still leaves plenty of image data to find faces in the image, but doing so takes less processing power. Note at this point that you keep the original image because this is what you'll use to display the found faces.

The cascade classifier returns a set of rectangle specifications as an upper-left point in x and y pixel position and a width and height of the rectangle from that origin. Remember, the classifier works on an image that is reduced by half, so the position, width, and height all need to be doubled when translated back to the original image. Before displaying the image, you'll draw the rectangles onto the original image. Figure 12.4 shows the camera output with multiple tracked faces.

FIGURE 12.4 Multiple faces identified in a captured frame.

Tracking a Face

From here, you can get to a fun project idea. The programs you've developed in this chapter can be combined with a skill you learned in Chapter 10, "Remote Monitoring and Data Collections." You can combine the use of two servos configured into something called a pan/tilt mechanism (SparkFun ROB-10335). The idea is simple: One servo moves the mechanism back and forth while the other servo tilts the platform up and down. Figure 12.5 shows this mechanism put together with two servos (SparkFun ROB-10333 and SparkFun ROB-09065).

You can combine this pan/tilt functionality with the face tracking capability to provide a platform that "targets" a face. This can be used for a variety of purposes, including moving the camera to keep it centered on a face. You can also keep the snapshot functionality from earlier. First, let's look at the hardware configuration. Figure 12.6 shows the breadboard configuration for our program.

FIGURE 12.5 A pan/tilt mechanism.

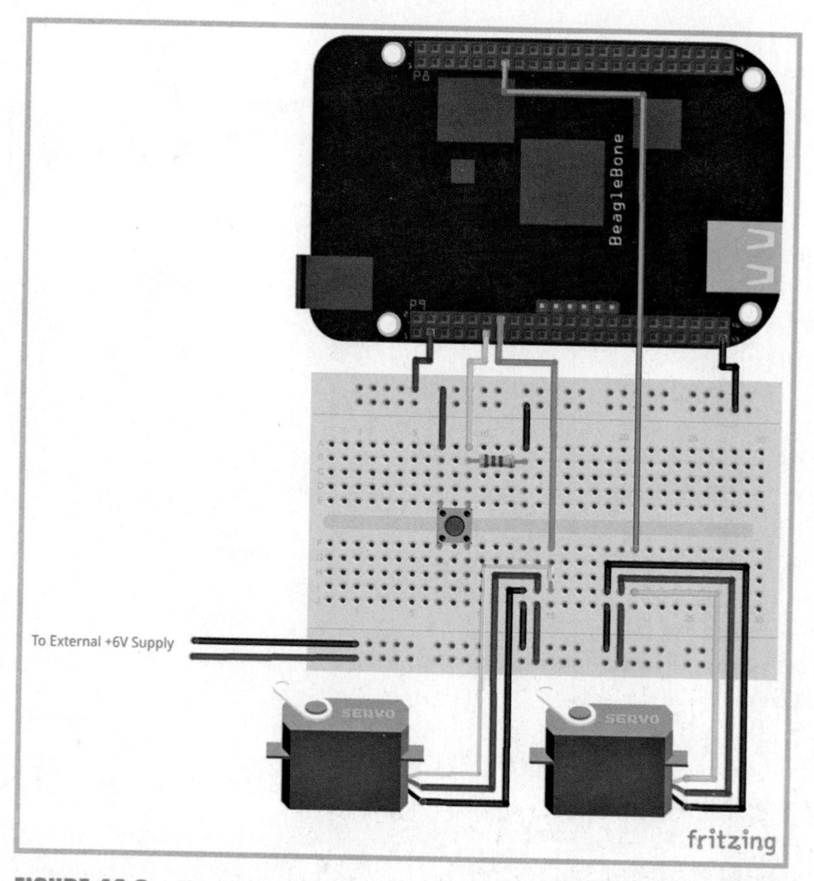

FIGURE 12.6 The advanced face tracker program's hardware.

For this program, to improve readability, you'll write your own library to drive the servos. This will take the basic servo functionality from Chapter 10 and encapsulate it in an object. This makes the main code, where you will drive the servos, easier to read. Listing 12.5 has the library with the `Servo` object definition.

LISTING 12.5 bbbservo.py

```
import Adafruit_BBIO.PWM as PWM


class Servo(object):
    '''Object representing a servo motor utilizing the Adafruit_BBIO
    library. This class defines an object to manipulate a
    servo motor on a BeagleBone Black (BBB). The class utilizes
    the PWM module of the Adafruit_BBIO library for pulse width
    manipulation.
```

```
Attributes:
    pin (string): The physical pin on the BBB (i.e. 'P9_14')
    servo_range (tuple): Max and min of servo range in degrees
    ppm_range (tuple): Max and min of the PPM pulse width in ms
    ppm_freq (number): Frequency of the PPM/PWM driver
    position (number): The position of the servo in degrees
    position_pulse_width (number): Position as pulse width
    position_duty_cycle (number): Position as a PWM duty cycle
    initialized (boolean): Object initialization status
'''
pin = ''
'''BBB PWM pin in use'''
servo_range = (0, 180)  # degrees
'''Range of the servo in degrees'''
ppm_range = (0.5, 2.5)
'''Range of the PPM pulse width for servo control in milliseconds'''
ppm_freq = 50
'''PWM/PPM Driver frequency'''
def __init__(self, pin, start, **kwargs):
    '''Initialize a Servo object.'''
    self.pin = pin
    for key in ('servo_range', 'ppm_range', 'ppm_freq'):
        if key in kwargs:
            setattr(self, key, kwargs[key])
    self.initialized = False
    self.position = start
@property
def position(self):
    '''Current servo position - assignment moves the servo'''
    return self._position
@position.setter
def position(self, value):
    if self.servo_range[0] <= value and value <= self.servo_range[1]:
        self._position = value
        if self.initialized:
            PWM.set_duty_cycle(self.pin, self._position_duty_cycle())
        else:
            PWM.start(self.pin, self._position_duty_cycle(),
                      self.ppm_freq)
            self.initialized = True
    else:
```

```
            message_string = 'Servo commanded to {}. Valid range {}.'
            raise Exception(message_string.format(value, self.servo_range))
    def cleanup(self):
        '''Execute steps to clean up the PWM hardware'''
        PWM.stop(self.pin)
        PWM.cleanup()
        self.initialized = False
    def _position_duty_cycle(self):
        '''Current servo position as a PWM duty cycle'''
        ppm_delta = self.ppm_range[1] - self.ppm_range[0]
        range_delta = self.servo_range[1] - self.servo_range[0]
        pw_per_degree = ppm_delta / float(range_delta)
        relative_position = self.position - self.servo_range[0]
        relative_pulse_width = relative_position * pw_per_degree
        absolute_pulse_width = relative_pulse_width + self.ppm_range[0]
        pulse_width_seconds = absolute_pulse_width / float(1000)
        return (pulse_width_seconds * self.ppm_freq) * 100
```

This code has a lot of documentation that you can use as an exercise to understand it in more depth. The attributes listed in the object documentation show how the basic functionality is abstracted and encapsulated. The library even imports the Adafruit_BBIO library. All you have to do in the main program is create two servo objects and use them. You can read a tracked position by reading the `.position` property, and by setting the `.position` property, you can move the servo. Note that servos don't provide position feedback to the controller, so the tracked position assumes the servo has achieved the commanded motions.

A difference from the previous program that tracks a face is that because you cannot track all of the faces simultaneously, the `for` loop differentiates between the first face returned, which is the chosen face to track, and the remaining faces. Also, you want to track the center of the "target," so you'll calculate the center of the face rectangle and mark it with a circle this time. A lot of calculations must take place before you start the primary loop to help reduce the number of calculations that take place inside the already processor-intensive loop. Listing 12.6 provides the source code, and Figure 12.7 shows a setup of the project.

LISTING 12.6 tracker.py

```
import Adafruit_BBIO.GPIO as GPIO
import bbbservo
import cv2
import time
from datetime import datetime
```

```
# Define BBB Pin Constants
BUTTON_PIN = 'P9_11'
PAN_PIN = 'P9_14'
TILT_PIN = 'P8_13'
# Field of view of the camera
FOV = 75
if __name__ == '__main__':
    try:
        # Configure the GUI
        window_name = 'Camera Tracker'
        cv2.namedWindow(window_name, cv2.WINDOW_NORMAL)
        # Configure the snapshot button
        GPIO.setup(BUTTON_PIN, GPIO.IN)
        GPIO.add_event_detect(BUTTON_PIN, GPIO.RISING)
        # Set up the identification cascade to use
        cascade = '/root/opencv/data/lbpcascades/lbpcascade_frontalface.xml'
        face_cascade = cv2.CascadeClassifier(cascade)
        # Define some basic camera frame properties and relationships
        camera = cv2.VideoCapture(-1)
        cam_frame_size = (camera.get(cv2.CAP_PROP_FRAME_WIDTH),
                          camera.get(cv2.CAP_PROP_FRAME_HEIGHT))
        cam_frame_center = (cam_frame_size[0] / 2,
                            cam_frame_size[1] / 2)
        pos_to_angle = ((FOV / 2) / cam_frame_center[0],
                        (FOV / 2) / cam_frame_center[1])
        # Configure and initialize out pan/tilt mechanism servos
        pan_servo = bbbservo.Servo('P9_14', 0,
                                   servo_range=(-90, 90),
                                   ppm_range=(0.4, 2.25))
        tilt_servo = bbbservo.Servo('P8_13', 0,
                                    servo_range=(-90, 90),
                                    ppm_range=(0.8, 2.25))
        time.sleep(1)  # Wait for servo motion to complete
        # Initialize a couple other parameters we will use
        last_found = time.time()
        while True:
            # Capture frame-by-frame
            ret, frame = camera.read()
            # Create a reduced copy of the frame and convert it to grey
            reduced = cv2.resize(frame, (0, 0), fx=0.5, fy=0.5)
```

```
            reduced_gray = cv2.cvtColor(reduced, cv2.COLOR_BGR2GRAY)
            # Find the faces in the reduced, grayscale, image
            faces = face_cascade.detectMultiScale(reduced_gray, 1.3, 2)
            for i, (x, y, w, h) in enumerate(faces):
                origin = (x * 2, y * 2)
                face_center = (origin[0] + w, origin[1] + h)
                if i == 0:
                    cv2.circle(frame, face_center, w, (0, 255, 0), 1)
                else:
                    cv2.circle(frame, face_center, w, (0, 0, 255), 1)
                center_delta = (cam_frame_center[0] - face_center[0],
                                cam_frame_center[1] - face_center[1])
                pan_servo.position = center_delta[0] * pos_to_angle[0]
                tilt_servo.position = center_delta[1] * pos_to_angle[1]
                last_found = time.time()
        # Otherwise, zero the platform if we haven't seen any faces
        # for a specified number of seconds
        else:
            if time.time() - last_found > 5:
                pan_servo.position = 0
                tilt_servo.position = 0
        # If the button has been pressed
        if GPIO.event_detected(BUTTON_PIN):
            # Grab the current time for a timestamp in the filename
            # and generate the file path & name
            snap_time = datetime.now().strftime('%Y-%m-%d_%H-%M-%S')
            output_file = 'snaps/snap_{}.jpeg'.format(snap_time)
            jpeg_settings = [int(cv2.IMWRITE_JPEG_QUALITY), 90]
            # Write the frame to the file
            cv2.imwrite(output_file, frame, jpeg_settings)
        # Display the resulting frame
        cv2.imshow(window_name, frame)
        cv2.waitKey(1)
except KeyboardInterrupt:
    camera.release()
    cv2.destroyAllWindows()
    pan_servo.position = 0
    tilt_servo.position = 0
    time.sleep(1)
    pan_servo.cleanup()
    tilt_servo.cleanup()
```

FIGURE 12.7 The finished tracker setup.

This code represents a great deal of progress from where you started at the beginning of your journey. You are truly using the power of the BeagleBone Black, pulling in information with the camera as a sensor, and moving a mechanism as a result. In the next chapter, we will look at more portable setup options by putting the BeagleBone Black in a car.

13

Sniffing Out Car Trouble

Small, embedded computers and microcontrollers are ubiquitous in many societies. When you look around a room, the number of devices you see that have some kind of controller or microprocessor is stunning. Unless you are driving a car that is now considered "classic," chances are your car has a computer in it, too. Generally called the Engine Control Module (ECM) or Engine Control Unit (ECU), this computer can do everything from monitoring systems in cars, to turning on the "Check Engine" light should there be a problem, to controlling the engine systems and making the car run. In this chapter, you explore how to use a BeagleBone Black to interface with a car's computer and get a look at some of this collected information.

Car Computers

When I was a young man in the late 1980s, I remember hearing about the car a friend's dad owned having a computer. I remember looking in the windows for something resembling a Commodore 64, TRS-80, or even the "portable computer" my dad would bring home from work occasionally. These were, at the time, what I thought of when I thought of a computer that was small and portable. Alas, I figured, it must have been hidden under the seats. I shrugged and went back to writing an application in QuickBasic to keep track of lawn-mowing customers and printing invoices for them.

Today, the idea of a car having a computer isn't a surprise to any of us. Microprocessors and microcontrollers are at the heart of almost every piece of technology in our lives. Apparently, some Amish are even starting to see technology creep into their lives—it is becoming just that common. A computer in a car isn't a new idea. Volkswagen started introducing an onboard computer with the capability to store and retrieve engine information in the late 1960s, and through my own fuzzy memory, I believe it was a 1973 Karmann Ghia that my friend's dad owned that I heard had the computer inside (see Figure 13.1). It was revolutionary at the time, but so were the onboard computers for the Apollo Command Module and Lunar Excursion Module. It was the early days of computers as part of the noise of life.

FIGURE 13.1 The Karmann Ghia in question. Thanks to Marshall Rouse for the photograph!

Over the years the use of computers in cars became more common among manufacturers around the world. For most cars owners, the computer simply turned on the "Check Engine" light, also known as the Malfunction Indicator Lamp (MIL), and the repair shops housed the necessary diagnostic computers to link in and read the information. As the technology grew, so did the needs for commonality and a standard. This standard, known as the On-Board Diagnostics (OBD), and the follow-on OBD-II, introduced a common way of communicating with a car's ECM. The OBD-II became a mandatory feature in the United States in the late 1990s, with Europe adopting a similar standard, the European OBD (EOBD), a couple years later. The standard includes a connector, electrical power and signal, a data bus, and data formatting.

There is an old saying that "rules are meant to be broken." There is a corollary to this in the world of computer standards that has been attributed to different people over the years, often to Admiral Grace Hopper: "The wonderful thing about standards is that there are so many of them to choose from." The 16-pin connector that is standard for OBD-II includes six pins for use of one of three different communications standards that are allowed in OBD-II and seven pins that different manufacturers can use as they want. The remaining three pins provide a connection to the car's battery voltage and ground. Figure 13.2 illustrates the OBD-II connector, and Table 13.1 explains the pin-out.

The standard includes a set of common modes and commands known as parameter IDs (PIDs). Keeping in mind what I said earlier about standards, you should note that not every mode and PID is supported by every car, but at least a few things are required to be in place. In fact, the PIDs that are publicly documented are just a core set; a car may have many more that it could respond to, but they are proprietary by manufacturer (and expensive to get a copy of). Because of this, we will focus on the subset that is readily accessible and start scanning our own cars for information.

FIGURE 13.2 The OBD-II standard connector.

TABLE 13.1 OBD-II Connector Pinout

Name		SAE J1850		Chassis Ground	Signal Ground	CAN	ISO 9141-1 & 14230	
Signal		+		GND	GND	High	K-Line	
Pin	1	2	3	4	5	6	7	8
Pin	9	10	11	12	13	14	15	16
Signal		-				Low	L-Line	V_{Batt}
Name		SAE J1850				CAN	ISO 9141-1 & 14230	Battery Voltage

Interfacing to the Car

The idea of dealing with the different standards and protocols may seem intimidating. This is why we are going to turn, once again, to our friends at SparkFun, who have created a board that can interface with the OBD-II system and make the interactions accessible over a serial UART connection, similar to what we used with the LCD screen in Chapter 7, "Expanding the Hardware Horizon." We actually need three parts. First, we need a cable that goes from the OBD-II interface to a 9-pin connector (SparkFun CAB-10087). Next, we need the OBD-to-UART board (SparkFun WIG-09555). Finally, because the UART output of the translator board is 5V, we need a logic-level converter to bring the signal down to a BeagleBone Black–friendly 3.3V for the receiver line and then boost it back to 5V on the transmitter line (BOB-12009). Figure 13.3 shows these parts laid out, whereas Figure 13.4 illustrates how to connect them all together.

NOTE

The Logic-Level Converter

An important item in this chapter is the logic-level converter. Remember, we are limited to 3.3V for the pins on the BeagleBone Black, with the exception of the analog inputs at 1.8V. Many devices, such as this SparkFun board, communicate at 5V. A logic-level converter such as this is helpful in bridging the gap between the two voltage standards.

FIGURE 13.3 Car connection parts.

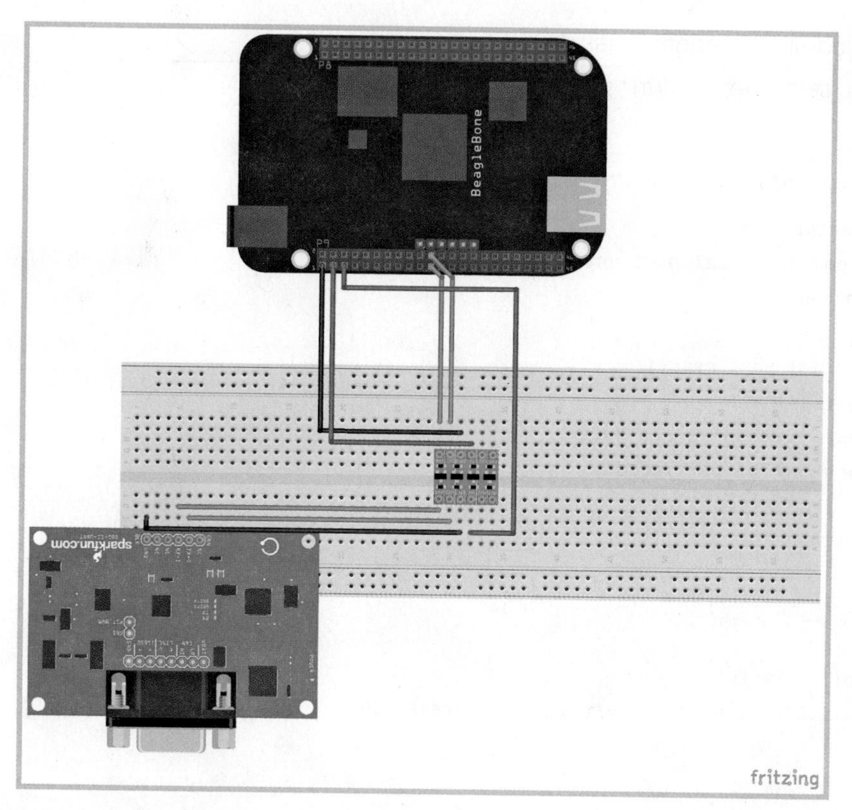

FIGURE 13.4 Connection diagram.

The two things we can talk to through this connection are the translation board itself and the car's computer. We will start with a basic Python script, shown in Listing 13.1, that connects using UART2 from the BeagleBone Black, as shown in the connection in Figure 13.4, and we'll send the command to reset. This ensures that we have basic communications working. Normally, in a Linux environment, a line ending is indicated with a linefeed ASCII control character (`'\n'`), ASCII character 10 (0x0A). In some environments, such as Windows, it is a combination of the carriage return (`'\r'`) character (ASCII 14, 0x0D) and line feed. In this board's environment, an end-of-line is indicated with a single carriage return. Also, we know the board is ready when we see the following character:

```
'>'
```

This is used as a control prompt by the board's communications.

At this point, you should note that the communications board needs to be connected to the OBD-II connector to operate, which means we need to be in the car. How you work on some of this prototyping in a car is up to you. I have a minivan where I was able to put the seats down and set up a work area in the back and communicate with the BeagleBone Black through the USB network connection we used in an earlier chapter. You can also connect a keyboard and drag a monitor out to your car. It is a matter of personal preference.

LISTING 13.1 communications_test.py

```
import Adafruit_BBIO.UART as UART
import serial


def obd_read(serial_port):
    '''Read from the serial port until the read prompt,
    ">", is encountered.'''
    response = ''
    a_byte = serial_port.read()
    while a_byte != '>':
        response = response + a_byte
        a_byte = serial_port.read()

    return response

if __name__ == '__main__':

    UART.setup('UART2')
    obd = serial.Serial(port='/dev/tty02', baudrate=9600)
    obd.open()

    obd.write('ATZ' + '\r')
    print(repr(obd_read(obd)))

    obd.close()
```

The board itself replies to two sets of commands prefixed with either `'AT'` or `'ST'`. Anything else it sees it assumes needs to be forwarded over the OBD to the car's ECM. The `'ATZ'` command sent in Listing 13.1 resets the serial interface. What you expect to see back is an echo of the command that was sent, some carriage return characters, information on the hardware, and the `'>'` prompt. Note that before you send the command string, you need to append the carriage return character.

The `obd_read` function defined in Listing 13.1 forms the core of communications you will see throughout the chapter. This reads in a response from the serial port one byte at a time and stores it all in a string until it encounters the `'>'` prompt. Because you know this means the board is now ready for the next command, you can consider that to be the end of the response. The `print` command actually prints the representation of the returned string. This is to ensure that you see all of the carriage return characters and any other control characters that are a part of the response. When the script is run, you should see the following:

```
debian@beaglebone:/root/bbb-primer/chapter13$ python connection_test.py
'ATZ\r\rELM327 v1.3a\r'
```

Note that your version information may differ from the example, but this is a good response. Now, to create an interactive terminal type application, a useful tool to test out some of the commands, you will move the communications into a class. This move into a class is similar to how you built a separate class for the servo control in the previous chapter. Create a starter version of the class, as shown in Listing 13.2, and expand it from there. For now, it will just provide the basic functionality shown in Listing 13.1.

LISTING 13.2 obd.py

```
import serial


class OBDUart(object):

    def __init__(self, port, baudrate, line_end='\r', prompt='>',
                 echo=True):
        self.port = port
        self.baudrate = baudrate
        self.line_end = line_end
        self.prompt = prompt
        self.echo = echo

        self.ser = serial.Serial(port=self.port, baudrate=self.baudrate)
        self.ser.open()

    def send(self, command):
        self.serial.write(command + self.line_end)

    def read(self):
        '''Read from the serial port until the prompt is encountered.'''
        response = ''
        a_byte = self.ser.read()
        while a_byte != self.prompt:
            response = response + a_byte
            a_byte = self.ser.read()

        return response

    def close(self):
        self.serial.close()
```

```
    def command(self, command):
        self.send(command)
        response = self.read()

        response = response.split(self.line_end)
        response = [line.strip() for line in response]

        if self.echo:
            if command not in response[0]:
                msg = 'OBDUart: Echo check failed.'
                raise Exception(msg.format(response[0]), command)
            else:
                response.pop(0)

        return response
```

By defining this class in a separate file, you can start abstracting the details of the implementation away in this class. It also allows you to use the interface in more complex ways. A couple items should be noted. The class method `command()` that's defined takes a command. This single method wraps the `send` and `read` commands. It also strips out any of the end-of-line characters and returns the response as a list of lines. If the command `echo` is turned on, the `command()` method checks to see if the command is in the first item in the response and raises an error (aka an Exception) to stop the program if it is not. This is the first level of error checking, which can save us headaches down the road.

By creating this class and saving it in a separate Python file (obd.py), you can see that the code from Listing 13.1 becomes the simple code shown in Listing 13.3.

LISTING 13.3 communications_test.py (revisited)

```
import Adafruit_BBIO.UART as UART
import obd


if __name__ == '__main__':

    UART.setup('UART2')
    obd_connection = obd.OBDUArt(port='/dev/tty02', baudrate=9600)

    print(repr(obd_connection.command('ATZ')))

    obd_connection.close()
```

Now you can create a slightly more complex program (fairly easily) that gives you interactive terminal-style access to the device. It is shown in Listing 13.4.

LISTING 13.4 interactive.py

```
import Adafruit_BBIO.UART as UART
import obd

if __name__ == '__main__':

    UART.setup("UART2")
    obd_con = obd.OBDUart(port='/dev/ttyO2', baudrate=9600)

    while True:
        cmd = raw_input('> ')
        if cmd == 'q':
            break
        else:
            response = obd_con.command(cmd)
            for msg in response:
                print msg

    obd_con.close()
```

This simple program we'll use to explore the OBD-II to UART device and even begin interacting with the car ECM itself! Granted, you can do the same thing with a terminal program over the serial connection, but with just these few lines of code you are already managing some of the underlying interfaces of the communications, such as line endings and checking for an expected response. Try sending a couple more commands to the board itself and then start talking to the car utilizing the Python code as follows:

```
debian@beaglebone:/root/bbb-primer/chapter13$ python interactive.py
>ATZ
ELM327 v1.3a
>ATSP0
OK
>ATDP
AUTO, ISO 15765-4 (CAN 11/500)
>STMFR
SparkFun Electronics
>ATRV
14.4V
```

You are starting to get good read-back from the OBD-II-to-UART communications board. The ATZ command returns information on the interface communications type. ATSP0 sets the data protocol to automatic; you ask the board to figure out the best way to talk to the car. You then interrogate which protocol it has decided to use with ATDP. In this case, it lets you know that it has auto-selected ISO 15765-4 (CAN 11/500), the CAN bus. If you look at the connectors in Table 13.1 and Figure 13.2, you can see that this means it is using pins 6 and 14 for communications with the car's ECM. STMFR replies with the manufacturer, SparkFun Electronics, and the battery voltage seen at the interface, interrogated with ATRV, is 14.4V.

Reading the Car's Status

So far, you've only communicated with the communications board, not the car itself. To start understanding how to communicate with the car's ECM, you need to understand the command structure. Finding some of this information is easy. Finding it consistently is a little more difficult. Most of this information comes from multiple resources. Although OBD-II is a standard, it isn't necessarily an open standard. In addition to the known commands (and how to interpret them) that we will discuss here, there are many additional commands that require expensive documents to access—and then the information cannot be reprinted or republished. Each manufacturer may have many additional command combinations, but those are not easy to find, and I actually could not find them for either my Honda or my Dodge.

In general, OBD-II commands have two parts: a mode and a parameter ID (PID). Generally, there is one byte for the mode and one byte for the PID. We will look only at a couple modes here, but I encourage exploration to see which modes and PIDs are available for your own vehicle. Google and Wikipedia will be your friend in this journey. The modes and PIDs are almost universally recognized by their hexadecimal representation, not their decimal representation, and we will keep that notation here.

Mode 0x01 represents a set of the current data in the vehicle. Starting with PID 0x00 in mode 0x01, every 0x20 PID responds with a binary representation, in 4 bytes, of the availability of the first 0x20 (0x01 to 0x20) PIDs. It seems safe to assume that Mode 0x01 PID 0x00 will always be available. PID 0x20 tells us which, if any, of the next 0x20 PIDs are available. You can use the interactive script you wrote to interrogate these PIDs to understand a little better. You send a command by just sending the mode followed by the PID all together. So for Mode 0x01 PID 0x00, we send the following:

```
>0100
41 20 80 07 E0 19
```

We received 6 bytes, but only the last 4 bytes tell us about PID availability. The first byte is always the mode requested plus 0x40, and the second byte represents the PID requested. So 0x41 and 0x20 let us know that this is a response to that PID request. The remaining bytes are the data for the response. We need a common way of interpreting the bytes, so we will stick with the standard used in the research for this chapter. We will name the bytes to which we are referring alphabetically, so the first byte is byte A, the second byte B, and so on. In a byte, we refer to a bit by its number from the *least significant bit* up, starting with 0.

You can see why this can be confusing. In this case, bit 7 of byte A represents the first Boolean value you want to check, but if this byte just represented a binary number, the LSB is in bit 0, which is actually the last bit of the byte. We will try to keep this all straight with a couple more examples. PID 0x0D has a simple 1-byte response, so let's check that one next. This PID replies with the vehicle's speed in kilometers per hour, so it is also a parameter that might be of interest.

```
>010D
41 0D 00
```

The 1-byte response is 0. Maybe this isn't the best first example, but it is a response that can be verified as accurate because the vehicle was stationary at the time! In this case, you are just interpreting byte A as an 8-bit, unsigned integer of, well, 0. Easy enough to understand. Another common parameter of interest is the tachometer, the meter that tells us how many revolutions per minute the engine is running. This is a 2-byte, or 16-bit, unsigned integer. The tachometer information is a 2-byte, or 16-bit, unsigned integer with a PID of 0x0C.

```
>010C
41 0C 09 98
```

These two bytes are separated by a space, but they are meant to be read together. Easy enough to check: The value 0x0998 is decimal 2456, which is the indicated RPM on the vehicle's dashboard as well. The engine was being revved at the time.

Interpreting the Data

Now you are getting a feel for how to talk to the car. Some responses require a lot more work to interpret. You also need to remember that what you are receiving from the computer is actually a character string representation of the numbers, not the numbers themselves.

Rather than four bytes, you actually see 11 bytes come across the connection. Table 13.2 shows the translation from the ASCII characters to the number values.

TABLE 13.2 OBD-II ASCII Value Interpretation

UART Hex	0x34	0x31	0x20	0x30	0x43	0x20	0x30	0x39	0x20	0x39	0x38
ASCII	'4'	'1'	[Space]	'0'	'C'	[Space]	'0'	'9'	[Space]	'9'	'8'
Result	0x41			0x0C			0x0998				
Decimal	65			12			2456				

Python makes this translation easy with a built-in function called `int()`. This function takes two parameters: a string representing an integer number, and the number of equivalent bits. So, for our data, in Python, the string `'0998'` can be converted into a number by using the following:

```
>>> int('0998', 16)
2456
```

So, here is what you want to do:

1. Create a `Car` object that takes an `OBDUart` object. This `Car` object will allow us to have an abstracted interface to the car and will initialized the `OBDUart` interface for us.
2. Inside the `Car` object, the byte string received, as a response, should be split into individual byte strings, each byte should be interpreted to an integer, and a new list of integer bytes created.
3. The first two bytes should be matched against the command sent and raise an exception as an error if there is a mismatch.
4. The remaining data bytes should be interpreted as necessary and returned to the user.

Listing 13.5 shows the start of a `Car` object that meets the previous objectives. You will put this object definition in the obd.py file.

LISTING 13.5 Car Object Definition (Within the obd.py File)

```
class Car(object):

    def __init__(self, obd_connection):
        self.obd_connection = obd_connection

        initialize_response = self.obd_connection.command('ATZ')
        if 'ELM' not in initialize_response:
            raise Exception('obd.Car.init: OBDUart Initilization Failed')

        autoset_interface = self.obd_connection.command('ATSP0')
        if 'OK' not in autoset_interface:
            raise Exception(
             'obd.Car.init: OBDUart Interface Autoset Failed')

    def raw_command(self, command):
        return self.obd_connection.command(command)

    def command(self, mode, pid):
        command_string = '{:0>2x}{:0>2x}'.format(mode, pid)
        response = self.raw_command.command(command_string)
```

```
        response = response.strip()
        response = response.split(' ')

        for byte in response:
            byte = int(byte.strip(), 16)

        response_mode = response[0] - 0x40
        response_pid = response[1]
        response_data = response[2:]

        if response_mode != mode or response_pid != pid:
            raise Exception(
              'obd.Car.command: Received unexpected Mode or PID')

        return response_data

    def speed(self):
        speed = self.command(0x01, 0x0D)
        return speed[0]

    def tachometer(self):
        tach = self.command(0x01, 0x0C)
        tach = (tach[0] << 8) | tach[1]
        return tach
```

In the initialization method, we reset the interface and set the protocol to auto-detect and check to make sure both commands worked. The `raw_command()` method allows the user of the library to still send whichever commands they want, as strings, to the system, as before.

The `command()` method that is part of this `Car` object takes a mode and PID as integers. This method first uses string formatting functionality to establish two fields, left justified as two digits, and to use the hexadecimal representation of the numbers in the field. This creates our command string. The command is sent through the OBD interface, and the response is received. The `strip()` string function removes any extra spaces from the ends of a string, which is then `split()` into a list of individual byte strings based on the space. Each byte is converted into an integer, and the first two bytes are compared against the commanded mode and PID to make sure we got a response we can trust.

Finally, the `speed()` and `tachometer()` methods are actual implementations of those functions in the car. For the `speed()` method, we can just return the first byte of the list because we know there is only one byte for the value. For the tachometer we need to combine the two bytes, so we shift the first byte up a full byte, back-filling with zeros, and then logically OR it with the second, lower byte. The result of this is a single 16-bit number representing the tachometer reading.

In just a few lines of code, we have the start of a simple interface to a car's computer from our BeagleBone Black, a highly portable computing device we can easily embed in a car. Let's take a look at a basic example. Back in Chapter 10, "Remote Monitoring and Data Collections," we built an environmental monitor that collects two parameters: temperature and light level. It automatically starts and saves the data to a file. We can do the same thing here very easily. Listing 13.6 shows an adapted version of the Chapter 10 program.

LISTING 13.6 car_monitor.py

```
#!/usr/bin/env python

import Adafruit_BBIO.UART as UART
import obd
import time
from datetime import datetime

SAMPLE_RATE = 2  # Hertz
TIMESTAMP_FORMAT = '%Y-%m-%d_%H-%M-%S'

if __name__ == '__main__':

    UART.setup("UART2")
    obd_con = obd.OBDUart(port='/dev/ttyO2', baudrate=9600)
    car = obd.Car(obd_con)

    start_time = datetime.now().strftime(TIMESTAMP_FORMAT)
    output_file = 'data_{}.jpeg'.format(start_time)

    try:
        with open(output_file, 'w') as f:
            while True:
                f.write('{}, {}'.format(car.speed(), car.tachometer()))
                time.sleep(1/SAMPLE_RATE)

    except KeyboardInterrupt:
        obd_con.close
```

NOTE

Important Note

Never fiddle with a project like this *while* you are driving. Have someone else do the driving, or set the connections up and don't touch them or worry about them while driving.

We can take this code, add in the work we did in Chapter 10 to make a program run on bootup, and we have a data logger we can easily take into a car. Because this is no longer a bench-top-only prototyping effort, a configuration is needed, such as the one shown in Figure 13.5, where a SparkFun ProtoCape has been attached and a breadboard has been placed on top of it for a more portable prototyping setup. This all is placed cleanly on the center console of a 2013 Dodge Grand Caravan.

In this chapter, you created an easy-to-use interface for connecting to a car computer. This can be used for a simple project, such as the data logger we created in this chapter, or something more complex, similar to KITT from *Knight Rider*. In the next chapter, you learn how to start tapping into the radio information floating in the air around you.

FIGURE 13.5 A more compact prototyping setup.

Ground Control to Major Beagle

In Chapter 13, "Sniffing Out Car Trouble," we discussed how computers, in some shape or form, are everywhere and how their ubiquity has increased since the middle of the last century. However, something else out there related to technology has been growing in ubiquity around us since the middle of the century *before* the last century. Radio energy moves as waves all around us at many different frequencies. In this chapter, you will mix together many of the things you have already learned so far with some knowledge of radio transmissions to use a BeagleBone Black capture radio signals.

Radio Data

The phrase "turn on the radio" refers to turning on a particular device that can tune to certain frequencies so you can listen to audio programming. Even today, streaming audio coming over a network interface is sometimes known as *Internet radio*. In reality, however, what we know as the AM and FM bands are a small part of the frequency range that encompasses what are considered radio frequencies, and a lot more than voice is transmitted over radio waves. Data of all kinds is transmitted—from audio, to television and other images, to satellite data. The spectrum allocations are shown in Figure 14.1, with the AM and FM radio bands highlighted.

So, what are radio waves? They are energy transferred via electromagnetic waves. Electricity and magnetism are always connected, and when electricity in a wire is alternating, it creates an electromagnetic wave. When that wave hits another piece of conductor, it induces electricity to start moving at the same rate. However, this is not a book on the physics of electricity and magnetism, which can get very complicated as you dive deeper into the subject. We simply use this as a fundamental concept moving forward in the chapter.

The term *spectrum* is used just like you think—like a rainbow. Light is a form of electromagnetic energy as well, and when you see a rainbow, you are looking at the intensity of all the different frequencies of light. When you are working with radio, this is often the best way to look at signal, as a plot where the horizontal axis represents increasing frequency and the vertical axis is the energy intensity at that point. As an example, take a look at Figure 14.2. This is the FM portion of the radio spectrum in the Baltimore/Washington D.C. area.

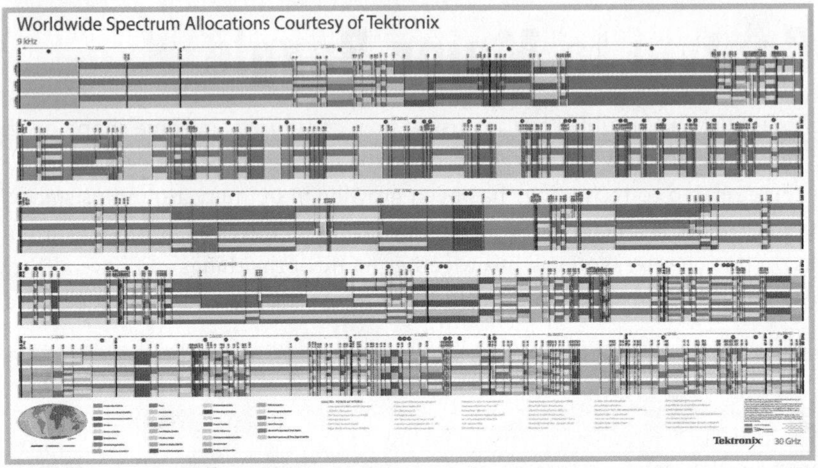

FIGURE 14.1 The electromagnetic spectrum from 9KHz to 30GHz (image courtesy of Tektonix).

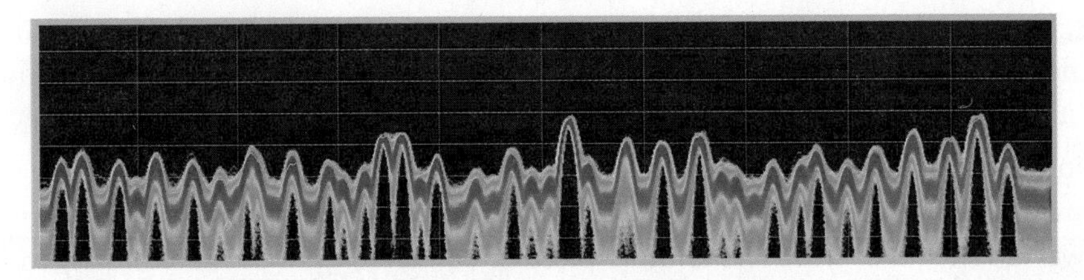

FIGURE 14.2 The full FM radio spectrum in the Baltimore/Washington D.C. area.

Figure 14.3 shows this part of the spectrum highlighted from Figure 14.1. As you can see, radio is only a very small portion of the spectrum. Note that each row in Figure 14.1 jumps an order of magnitude as to how much spectrum it covers. Compared to the 3GHz to 30GHz on the bottom line, this is a very small section of the spectrum.

Each peak in Figure 14.2 represents a different radio station, with the center of the peak at the dial frequency. For example, let's say we want to tune to a radio station that identifies as 90.1. This means that the center frequency for the transmission is at 90,100,000 Hertz, or 90.1MHz. This central frequency is generated by a very high-power transmitter oscillating an electrical current at that frequency. Information, including sound in the case of broadcast radio, is then added by modulating this frequency. All of that information is contained within 200KHz of the spectrum. This allows for all the information to be captured without overlapping the stations.

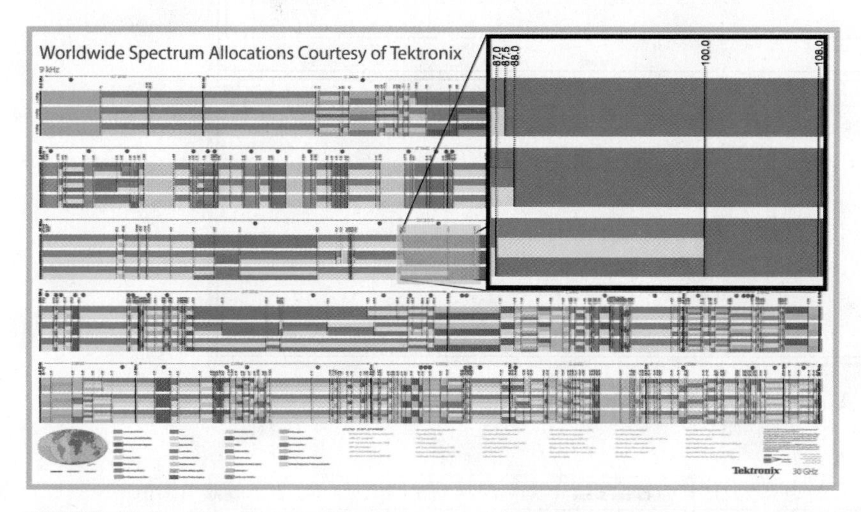

FIGURE 14.3 The FM radio spectrum highlighted (image courtesy of Tektronix).

You're already familiar with the terms for two forms of modulation: amplitude modulation (AM) and frequency modulation (FM). The term *modulating* means that some aspect of this carrier signal is changed based on the input. In amplitude modulation, the amplitude of the carrier is changed by the characteristics of the signal. In frequency modulation, the frequency is shifted with respect to the frequency of the input signal. So with amplitude modulation, the signal will rise and fall very rapidly on the spectrum as it is fluctuating with the signal. With frequency modulation, the frequency will shift rapidly around the carrier. Figures 14.4 through 14.6 show a signal undergoing amplitude modulation. Figure 14.4 shows the input signal (A above middle C on a piano) and the resulting spectrum clearly in the audio range from 0Hz to around 22KHz. If you look at this with a sharp eye, you can see that the piano is in need of tuning.

Figure 14.5 shows the carrier signal, a simple sine wave with a spectrum showing the carrier frequency around 500KHz.

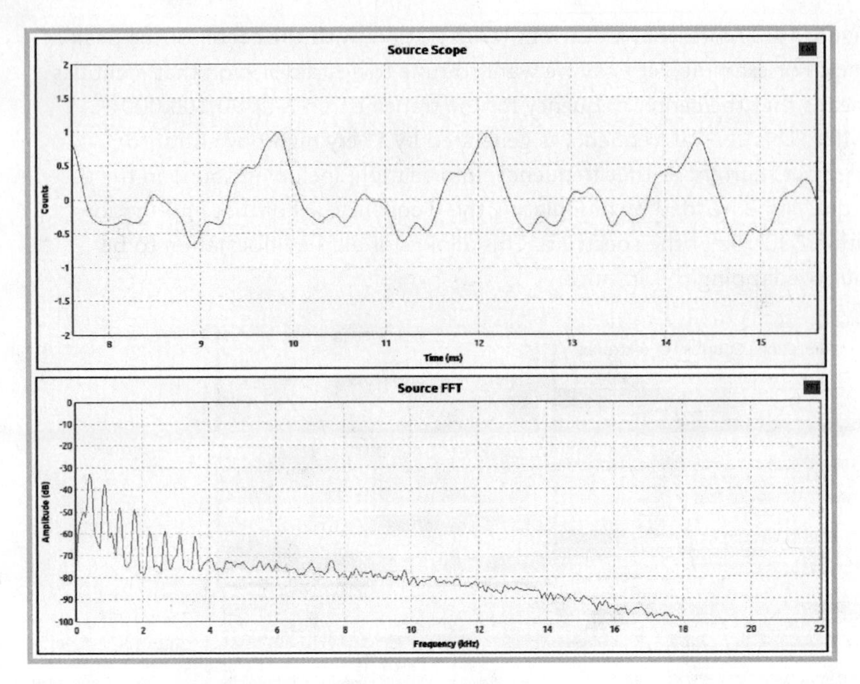

FIGURE 14.4 A above middle C on the piano.

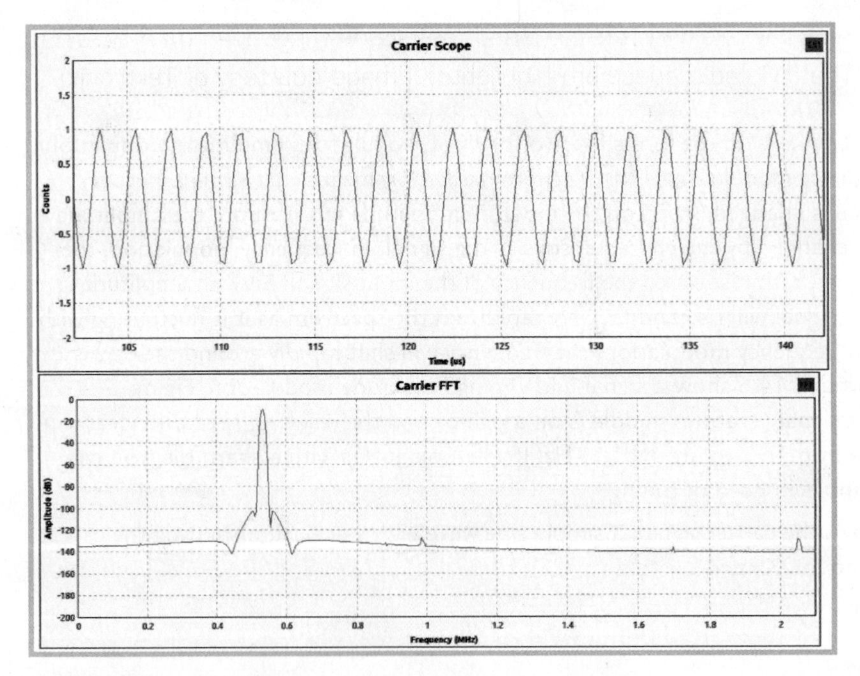

FIGURE 14.5 A 500KHz carrier signal.

Finally, Figure 14.6 shows the amplitude modulated signal, where the original input waveform is encoded into the changing amplitude of the carrier waveform.

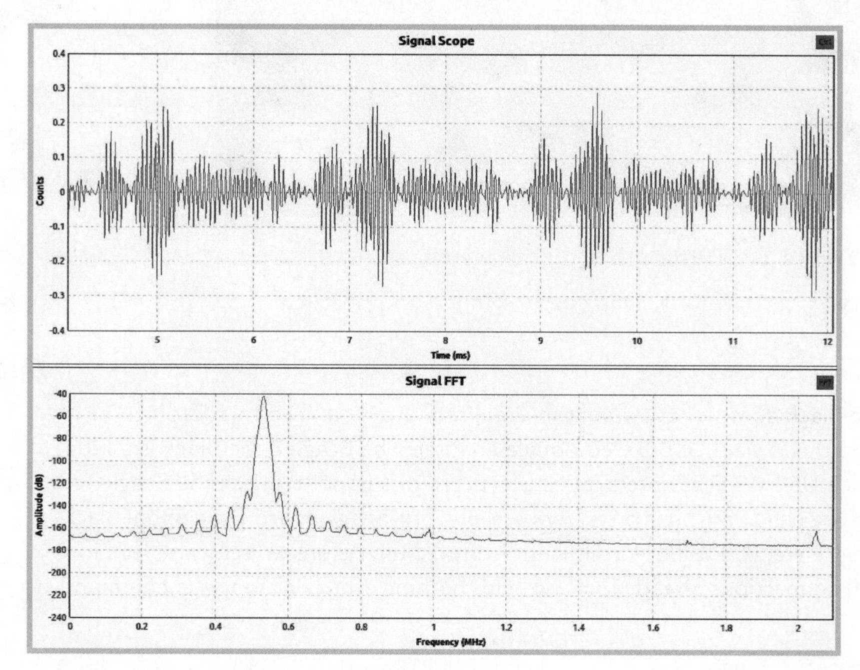

FIGURE 14.6 A above middle C on the piano, amplitude modulated on a 500KHz carrier.

Of course, you can encode many things in a signal beyond a voice, such as data. With data, you can also modulate by changing the phase of the signal, which is known as *phase modulation* or *phase shift modulation* (PSK). We won't cover the details here, but PSK is often used in data streams. The two data streams you are likely most familiar with are Wi-Fi and Bluetooth. These have become common in our world, and the streams are all around us. Sitting in the 2GHz band and 5GHz band, these signals can move a lot of data very quickly. Figure 14.7 shows Wi-Fi channel 8 in the 2.4GHz band. Notice that whereas radio is using 200KHz of spectrum, Wi-Fi is using 22MHz. Our first venture into radio using our BeagleBone Black is to add a Wi-Fi connection.

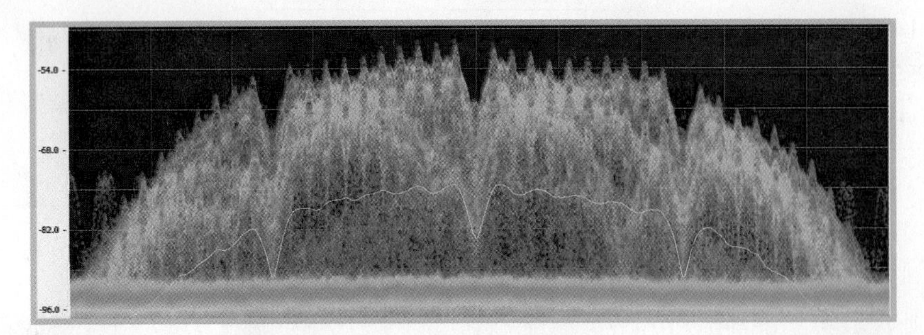

FIGURE 14.7 Wi-Fi spectrum.

WiFi

The BeagleBone Black doesn't have a built in Wi-Fi radio, so you need to add one. The easiest way to do this is via a USB-based dongle adapter. The BeagleBone Black is a little picky about which USB Wi-Fi adapters it will accept, so it is good to check the BeagleBone Black section of the elinux.org site (http://www.elinux.org/Beagleboard:BeagleBoneBlack#WIFI_Adapters). Here, you will use a TP-Link TL-WN721N. Before we set it up, let's look at what the USB connections already look like with nothing attached using the command `lsusb`:

```
root@beaglebone:~# lsusb
Bus 001 Device 001: ID 1d6b:0002 Linux Foundation 2.0 root hub
Bus 002 Device 001: ID 1d6b:0002 Linux Foundation 2.0 root hub
```

These are just the basic devices available on the BeagleBone Black. Now, you can insert the Wi-Fi adapter and run the `lsusb` command again:

```
root@beaglebone:~# lsusb
Bus 001 Device 002: ID 0cf3:9271 Atheros Communications, Inc.
AR9271 802.11n
Bus 001 Device 001: ID 1d6b:0002 Linux Foundation 2.0 root hub
Bus 002 Device 001: ID 1d6b:0002 Linux Foundation 2.0 root hub
```

You can see our 801.11n device now recognized on the USB system, and we should check to see if it is seen as a Wi-Fi network device. We can do this by using the `ifconfig` command with the `-a` option to see all interfaces, as shown here:

```
root@beaglebone:~# ifconfig -a
eth0      Link encap:Ethernet  HWaddr 7c:66:9d:58:bd:41
          inet addr:192.168.1.153  Bcast:192.168.1.255  Mask:255
          .255.255.0
          inet6 addr: fe80::7e66:9dff:fe58:bd41/64 Scope:Link
          UP BROADCAST RUNNING MULTICAST  MTU:1500  Metric:1
```

```
          RX packets:3359 errors:0 dropped:2 overruns:0 frame:0
          TX packets:314 errors:0 dropped:0 overruns:0 carrier:0
          collisions:0 txqueuelen:1000
          RX bytes:582966 (569.3 KiB)  TX bytes:44640 (43.5 KiB)
          Interrupt:40

lo        Link encap:Local Loopback
          inet addr:127.0.0.1  Mask:255.0.0.0
          inet6 addr: ::1/128 Scope:Host
          UP LOOPBACK RUNNING  MTU:65536  Metric:1
          RX packets:0 errors:0 dropped:0 overruns:0 frame:0
          TX packets:0 errors:0 dropped:0 overruns:0 carrier:0
          collisions:0 txqueuelen:0
          RX bytes:0 (0.0 B)  TX bytes:0 (0.0 B)

usb0      Link encap:Ethernet  HWaddr 92:4a:d1:2f:2e:75
          inet addr:192.168.7.2  Bcast:192.168.7.3  Mask:255.255.255.252
          UP BROADCAST MULTICAST  MTU:1500  Metric:1
          RX packets:0 errors:0 dropped:0 overruns:0 frame:0
          TX packets:0 errors:0 dropped:0 overruns:0 carrier:0
          collisions:0 txqueuelen:1000
          RX bytes:0 (0.0 B)  TX bytes:0 (0.0 B)

wlan0     Link encap:Ethernet  HWaddr c4:6e:1f:1d:7e:a2
          BROADCAST MULTICAST  MTU:1500  Metric:1
          RX packets:0 errors:0 dropped:0 overruns:0 frame:0
          TX packets:0 errors:0 dropped:0 overruns:0 carrier:0
          collisions:0 txqueuelen:1000
          RX bytes:0 (0.0 B)  TX bytes:0 (0.0 B)
```

The Wi-Fi adapter is the device listed as “wlan0,” and the hardware address is recognized. Great news! Now you need to tell it how establish a Wi-Fi connection. Edit the file at `/etc/network/interfaces`. In that file, you will find some lines under `WiFi Example`—in particular the lines labeled `wpa-ssid` and `wpa-psk`. Here, `wpa-ssid` shows the name of the wireless network and `wpa-psk` shows the Wi-Fi password. Filled in for the test network, this section appears as follows:

```
# WiFi Example
auto wlan0
iface wlan0 inet dhcp
    wpa-ssid "MilkyWay"
    wpa-psk  "************"
```

Note that the password is not actually a series of asterisks; this is used to protect the connected network's security. The password goes there as plain text. If you reboot the BeagleBone Black with everything configured properly, you will see that wlan0 is now populated with an IP address (`inet addr`):

```
wlan0     Link encap:Ethernet  HWaddr c4:6e:1f:1d:7e:a2
          inet addr:192.168.1.165  Bcast:192.168.1.255  Mask:255
          .255.255.0
          inet6 addr: fe80::c66e:1fff:fe1d:7ea2/64 Scope:Link
          UP BROADCAST RUNNING MULTICAST  MTU:1500  Metric:1
          RX packets:1748 errors:0 dropped:0 overruns:0 frame:0
          TX packets:537 errors:0 dropped:0 overruns:0 carrier:0
          collisions:0 txqueuelen:1000
          RX bytes:280547 (273.9 KiB)  TX bytes:70158 (68.5 KiB)
```

You are now successfully connected to Wi-Fi! You can disconnect the Ethernet cable or USB network cable to the computer. With just power and the adapter, you have a network-connected computer you can place anywhere in Wi-Fi range.

Software Defined Radio

Beyond WiFi, you can explore other radio options with something called a software-defined radio (SDR). The idea behind an SDR is simple: Rather than having dedicated hardware that processes the analog signal from radio, the signal is sampled (just like the analog-to-digital inputs on the BeagleBone Black used in a previous chapter) and then processed as data blocks. This means that the form of demodulation, where you recover the input signal from the carrier, can be changed quickly and a lot of other processing can be done easily with no hardware changing. A normal assumption would be that getting your hands on an SDR is an expensive process. However, there is a very cheap way to get into receiving signals with a software-defined radio.

A small USB device series is available that is designed to receive digital television signals. These relatively cheap devices cost approximately $20 and are easy to order. The chip utilized in these devices is very robust, and it didn't take long for someone to figure out that there is low-level access to this chip over the USB connection and that what was once just a TV tuner can now be a very useful SDR. The chipset is the RTL2832, so this kind of SDR is known as an RTL-SDR. A personal favorite of mine is available from Adafruit (Product ID 1497) and is shown in Figure 14.8. Let's take a look at connecting one to the BeagleBone Black and receiving a basic radio station.

FIGURE 14.8 RTL-SDR device.

First, you need to ensure that the device has adequate power. Since you will connect Wi-Fi and this radio device at the same time, you'll want to use a USB hub that has its own power plug. You can plug it in to the wall, connect it to the BeagleBone Black, and then connect the Wi-Fi dongle and RTL-SDR, as shown in Figure 14.9.

FIGURE 14.9 BeagleBone Black with a connected USB hub for Wi-Fi and RTL-SDR.

As when you installed the Wi-Fi dongle, you can run `lsusb` to see the device listed, as shown next:

```
root@new-host-5:~# lsusb
Bus 001 Device 002: ID 05e3:0607 Genesys Logic, Inc. Logitech G110 Hub
Bus 001 Device 001: ID 1d6b:0002 Linux Foundation 2.0 root hub
Bus 002 Device 001: ID 1d6b:0002 Linux Foundation 2.0 root hub
Bus 001 Device 003: ID 05e3:0607 Genesys Logic, Inc. Logitech G110 Hub
Bus 001 Device 004: ID 0bda:2832 Realtek Semiconductor Corp. RTL2832U
DVB-T
Bus 001 Device 005: ID 0cf3:9271 Atheros Communications, Inc. AR9271
802.11n
```

From this command output, you can see the list of devices you saw before plus the RTL2832U device. This is a good sign because the BeagleBone Black will have no problem talking to the device.

Grabbing Libraries with Git

Next, you need to dip into the pool of open-source libraries again and install the software to operate the radio. You are actually going to clone the source libraries from a git repository. Git is known as a revision control system (RCS). In an RCS, a user checks files in and out, depending on how they are being developed. This allows for a group of developers to work on multiple versions simultaneously. It can be used for large and complex projects and small, personal development projects alike. A site called GitHub (http://github.com) is a well-known repository for git-managed projects, but the source you need is hosted on a server at osmocom.org. The git software should be included with the BeagleBone Black distribution, but if you run the `git` command and the command is not found, it can be installed with the apt system, as follows:

```
root@new-host-5:~# apt-get update && apt-get upgrade
root@new-host-5:~# apt-get install git
```

You need a library to handle USB interactions with the RTL-SDR hardware, so you need to make sure you install the libusb-1.0 library:

```
root@new-host-5:~# apt-get install libusb-1.0
```

Once you have the `git` command at your disposal and the USB library, you can clone the repository (*clone* is a specific term here) and move into the created directory. You will then create a build directory and move into that directory. This is shown next:

```
root@new-host-5:~# git clone git://git.osmocom.org/rtl-sdr.git
Cloning into 'rtl-sdr'...
remote: Counting objects: 1587, done.
remote: Compressing objects: 100% (681/681), done.
remote: Total 1587 (delta 1160), reused 1213 (delta 898)
Receiving objects: 100% (1587/1587), 341.27 KiB ¦ 231 KiB/s, done.
Resolving deltas: 100% (1160/1160), done.
root@new-host-5:~# cd rtl-sdr
root@new-host-5:~/rtl-sdr# mkdir build
root@new-host-5:~/rtl-sdr# cd build
```

Next, you'll run a couple utilities that help configure the source files to build the software. The `-DINSTALL_UDEV_RULES` option ensures that users other than root (debian, for example) can use these tools. You will configure using a utility called `cmake`:

```
root@new-host-5:~/rtl-sdr/build# cmake ../ -DINSTALL_UDEV_RULES=ON
-- Build type not specified: defaulting to release.
```

```
-- Extracting version information from git describe...
-- checking for module 'libusb-1.0'
--   found libusb-1.0, version 1.0.11
-- Found libusb-1.0: /usr/include/libusb-1.0, /usr/lib/arm-linux-
gnueabihf/libusb-1.0.so
-- Udev rules not being installed, install them with -DINSTALL_
UDEV_RULES=ON
-- Building with kernel driver detaching disabled, use -DDETACH_
KERNEL_DRIVER=ON to enable
-- Building for version: v0.5.3-6-gd447 / 0.5git
-- Using install prefix: /usr/local
-- Configuring done
-- Generating done
-- Build files have been written to: /root/rtl-sdr/build./conf
igure --enable-driver-detach
```

The cmake command checks for required dependencies for the build process and configures the compiling process. Next, you build and install the software. The make command will take only a couple minutes to perform the build. (Note that the actual output from these commands has not been replicated here, so you should check for any error messages in the output.)

```
root@new-host-5:~/rtl-sdr/build# make
root@new-host-5:~/rtl-sdr/build# make install
root@new-host-5:~/rtl-sdr/build# ldconfig
```

Radio Testing

If all goes well, you now have a set of software and associated libraries you can use for operating the radio hardware. You can test it using the rtl_test command. In this case, the -t parameter tells the test program to check the tuning hardware of the device. Here you should note that these radio devices are mildly sensitive to temperature changes. A good rule of thumb is to set your device up, power it on, and then wait about 20 minutes or so for the device to get to a stable operating temperature. The more sensitive the measurement you need, the more important this will be.

```
root@new-host-5:~# rtl_test -t
Found 1 device(s):
  0:  Realtek, RTL2838UHIDIR, SN: 00000001

Using device 0: Generic RTL2832U OEM
Found Rafael Micro R820T tuner
Supported gain values (29): 0.0 0.9 1.4 2.7 3.7 7.7 8.7
```

```
12.5 14.4 15.7 16.6 19.7 20.7 22.9 25.4 28.0 29.7 32.8
33.8 36.4 37.2 38.6 40.2 42.1 43.4 43.9 44.5 48.0 49.6
Sampling at 2048000 S/s.
No E4000 tuner found, aborting.
```

What you can see from this output is that the device in use utilizes a Rafael Micro R820T tuner and supports 29 values for the device gain (these values are listed). *Gain* is the amount the input signal is artificially boosted. However, in this case, the signal, as well as any noise, are boosted together, so a higher gain will not always provide a better signal data.

Back in Chapter 9, " Interacting With Your World Part 1: Sensors" we discussed the fact that we would revisit the sample rate in the future. Well, we've reached the future. A good sample rate that works for most RTL-SDR devices is 2.048MHz, which is actually the rate used in the preceding `rtl_test` example. Due to the way these devices tune, this actually represents the bandwidth over which we can look. Figure 14.10 shows the spectrum of five FM radio stations, with a single station highlighted. Different signal types have different bandwidth needs. A typical FM commercial radio station broadcast uses 200KHz of bandwidth, or what is known as wideband FM (WBFM). Two other, lower-power stations are on either side of the highlighted station, and a carrier tone is seen on the leftmost station.

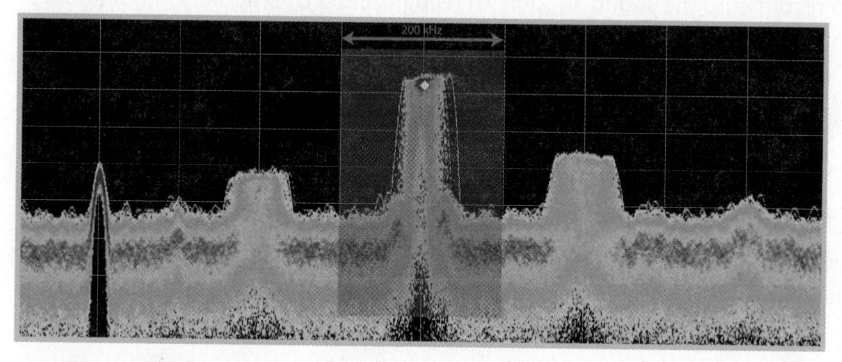

FIGURE 14.10 An FM radio station with the bandwidth illustrated.

A sample rate of 2.048MHz means that we are sampling 2.048MHz of signal centered on the tuned frequency. For example, if we tune to the radio station at 99.1MHz with a sample rate of 2.048MHz, we will sample, centered on 99.1MHz, with 1.024MHz on either side included.

You will test the new system out by building a simple FM radio device. Generally, the RTL-SDR device comes with an antenna you can connect, and this antenna does an okay job of receiving FM radio. When you build the RTL-SDR software, one of the utilities created, `rtl_fm`, will tune to a radio station. This command tunes to the station, samples at the specified bandwidth, demodulates to pull the original signal out of the frequency modulated signal, and provides an audio output.

If the BeagleBone Black is connected to a video monitor that supports audio-out over HDMI, then you can use that audio output. Another option is to use a USB sound device. Many such devices are available, and, depending on the audio quality desired, you don't need to spend a lot of money. The USB Audio Adapter at Adafruit (Product ID: 1475) is the device used in this example. First, however, you need to turn off HDMI output. The steps outlined here are also the steps you would use if you want to disable HDMI in general to make the HDMI pins on the connectors available for other uses. You can use the `aplay` command to see which audio devices are present in the system:

```
root@new-host-5:~# aplay -l
**** List of PLAYBACK Hardware Devices ****
card 0: Black [TI BeagleBone Black], device 0: HDMI nxp-hdmi-hifi-0 []
  Subdevices: 1/1
  Subdevice #0: subdevice #0
```

To disable this device, you need to modify a file in the system boot configuration, `/boot/uboot/uEnv.txt`, by adding the following line or uncommenting it, as is the case in the Debian release as of this writing:

```
cape_disable=capemgr.disable_partno=BB-BONELT-HDMI,BB-BONELT-HDMIN
```

You also need to reconfigure the sound support to recognize the USB audio as audio device zero. For this, you need to modify another configuration file, `/etc/modprobe.d/alsa-base.conf`. In this file, you can see the following line:

```
options snd-usb-audio index=-2
```

You need to modify this line as follows:

```
options snd-usb-audio index=0
```

At this point, you can connect your USB audio output as well and then reboot the BeagleBone Black. Now, if you execute the `aplay` command, you can see the USB audio device, as shown here:

```
root@beaglebone:~# aplay -l
**** List of PLAYBACK Hardware Devices ****
card 0: Device [USB PnP Sound Device], device 0: USB Audio [USB Audio]
  Subdevices: 1/1
  Subdevice #0: subdevice #0
```

You are now ready to tune the radio to a station. As mentioned previously, you will use the `rtl_fm` utility to help. For this example, we will tune to 99.1MHz, which happens to be the strongest signal in my area. You should update this to a strong station in your area when executing this example.

```
root@beaglebone:~# rtl_fm -f 99.1e6 -M wbfm -r 48000 - | \
aplay -r 48k -f S16_LE
```

```
Found 1 device(s):
  0:  Realtek, RTL2838UHIDIR, SN: 00000001

Using device 0: Generic RTL2832U OEM
Found Rafael Micro R820T tuner
Tuner gain set to automatic.
Tuned to 99416000 Hz.
Oversampling input by: 6x.
Oversampling output by: 1x.
Buffer size: 6.83ms
Sampling at 1200000 S/s.
Output at 200000 Hz.
Playing raw data 'stdin' : Signed 16 bit Little Endian,
Rate 48000 Hz, Mono
underrun!!! (at least 2067591660.491 ms long)
underrun!!! (at least 2067591660.487 ms long)
```

Let's take apart the parameters for this command to understand better what is occurring here. First, the `-f` parameter sets the desired frequency in Hertz (in this case, `99.1e6`, which is engineering notation for 99,100,000 Hertz, or 99.1MHz). The `-M` parameter sets the demodulation scheme to use. You are using wideband FM, or `wbfm`, as the parameter. Finally, `-r` indicates that the output audio rate should be 48KHz. You pipe all this to the `aplay` command, where you have the input set to match the output rate of the `rtl_fm` command (48KHz) and set the format of the data coming out of the `rtl_fm` command.

Calibrating the Radio

With these settings, you get a good signal, but there is one error you can correct to improve the signal slightly. As you can guess from the price, these devices are made fairly cheaply and the quality control is not the same as it would be for more expensive devices. For most purposes this isn't a big deal, save for one measurement, which is the frequency of the oscillator on the device itself. This can vary by a good amount from device to device; however, it can be calibrated and accounted for. The error is measured in parts per million, commonly referred to as *ppm*. A great utility has been created, called Kalibrate, that looks for markers in the cell phone GSM signals floating around and utilizes a frequency-correction signal buried in those channels to determine the error on the device. First, to install this software we will need to add a library:

```
root@beaglebone:~# apt-get install libfftw3-bin
```

Next, you need to clone the program to your BeagleBone Black from the git repository:

```
root@beaglebone:~# git clone git://github.com/steve-m/kalibrate-rtl.git
```

```
Cloning into 'kalibrate-rtl'...
remote: Counting objects: 85, done.
remote: Total 85 (delta 0), reused 0 (delta 0), pack-reused 85
Receiving objects: 100% (85/85), 31.60 KiB, done.
Resolving deltas: 100% (55/55), done.

root@beaglebone:~# cd kalibrate-rtl
```

You then need to execute the build process, which is a little different from the build processes we've used in the past:

```
root@beaglebone:~/kalibrate-rtl# ./bootstrap && CXXFLAGS='-W -Wall -03'

root@beaglebone:~/kalibrate-rtl# ./configure

root@beaglebone:~/kalibrate-rtl# make

root@beaglebone:~/kalibrate-rtl# make install
```

Now you can run the Kalibrate program. First, you need to look for valid GSM signals in your area. In the United States, this is in the GSM850 band.

```
root@beaglebone:~# kal -s GSM850
Found 1 device(s):
  0:  Generic RTL2832U OEM

Using device 0: Generic RTL2832U OEM
Found Rafael Micro R820T tuner
Exact sample rate is: 270833.002142 Hz
kal: Scanning for GSM-850 base stations.
GSM-850:
        chan: 180 (879.6MHz - 11.001kHz)        power: 39391.57
        chan: 227 (889.0MHz + 36.021kHz)        power: 53096.40
```

In this demonstration, only two channels are found: channels 180 and 227. The rig can be moved to other locations to look for better signals, which might result in more channels being found. You can also find more channels with the -g option and increasing the gain with one of the gain settings found from executing `rtl_test -t`. We will go ahead and calibrate on those channels. The following results demonstrate the output with channel 180:

```
root@beaglebone:~# kal -c 180
Found 1 device(s):
  0:  Generic RTL2832U OEM
```

```
Using device 0: Generic RTL2832U OEM
Found Rafael Micro R820T tuner
Exact sample rate is: 270833.002142 Hz
kal: Calculating clock frequency offset.
Using GSM-850 channel 180 (879.6MHz)
average         [min, max]        (range, stddev)
- 10.815kHz             [-10880, -10773]        (106, 27.839817)
overruns: 0
not found: 5
average absolute error: 12.296 ppm
```

This receives a ppm error of 12.296. If you run the command a couple times on a couple different channels, you will see an average calculated error. In this case, the error is seen around 12 ppm, so you will use that to calibrate out the frequency error. Remember, this program is telling us what the error is, so to calibrate it out we will need to negate it, so use –12 in this case. Otherwise, you end up doubling the error. To include this with the FM radio application, change the call to the following:

```
root@beaglebone:~# rtl_fm -f 99.1e6 -M wbfm -r 48000 \
-p -12 - ¦ aplay -r 48k -f S16_LE
```

Listening to Aviation Data

So far, with the SDR, we have looked at FM radio and GSM signals for calibration, but how about something a little more fun? We can build a system to tell us which aircraft are in the area using a communications protocol called ADS-B.

ADS-B runs at a frequency of 1.090GHz and is the communications protocol used by air traffic control to determine which planes are at which positions in the sky. It is a simple protocol, and Figure 14.11 shows an image of the ADS-B frequency activity. This activity is right in the range of our RTL-SDR device, so we can try and listen in as well. To start, you'll install another utility, dump1090. As with the last couple utilities you've installed, you clone a git repository, like so:

```
root@beaglebone:~# git clone git://github.com/antirez/dump1090.git
Cloning into 'dump1090'...
remote: Counting objects: 261, done.
remote: Total 261 (delta 0), reused 0 (delta 0), pack-reused 261
Receiving objects: 100% (261/261), 598.38 KiB, done.
Resolving deltas: 100% (139/139), done.
root@beaglebone:~# cd dump1090
```

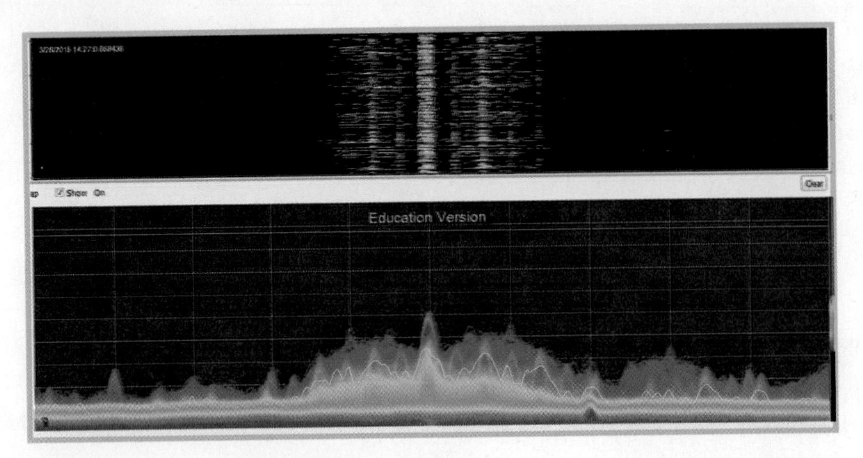

FIGURE 14.11 Aircraft ADS-B activity.

Then, for dump1090, simply run `make` to get a dump1090 executable. You can run the executable in this directory or copy it to a system path to make it available everywhere:

```
root@beaglebone:~/dump1090# make
cc -O2 -g -Wall -W -I/usr/local/include/ \
-I/usr/include/libusb-1.0    -c dump1090.c
cc -O2 -g -Wall -W -I/usr/local/include/ \
-I/usr/include/libusb-1.0    -c anet.c
cc -g -o dump1090 dump1090.o anet.o \
-L/usr/local/lib -lrtlsdr -lusb-1.0    -lpthread -lm
root@beaglebone:~/dump1090# cp dump1090 /usr/local/bin/
root@beaglebone:~# cd ~
```

Believe it or not, you are now ready to look at information from aircraft flying overhead. Let's see if we can get any data without any further work:

```
root@beaglebone:~# dump1090 --interactive

Hex     Flight    Altitude  Speed   Lat     Lon     Track  Messages Seen .
-------------------------------------------------------------------------------
ad1697            11350     0       0.000   0.000   0      193          0 sec
```

Straight off the bat, without any move in the lab setup, we can see one plane flying overhead at 11, 350 feet. What if we move to a location with more signal available? Just moving the setup to better position yields improved results, as shown here:

```
Hex     Flight    Altitude  Speed   Lat     Lon     Track  Messages Seen .
-------------------------------------------------------------------------------
aae960            37975     0       0.000   0.000   0      14           1 sec
a0a9f0            36000     0       0.000   0.000   0      4            1 sec
ada163            12750     0       0.000   0.000   0      27           1 sec
```

```
ab99df          28000     0       0.000   0.000  0      22          1 sec
a3c5ef          9875      0       0.000   0.000  0      95          0 sec
a53ba4          6925      210     39.197  -76.82 283    112         1 sec
aa435b          34025     0       0.000   0.000  0      49          1 sec
a8996b          34000     0       0.000   0.000  0      86          0 sec
```

In this case, I moved the entire rig (the BeagleBone Black, USB hub and devices, and antenna) to the window of an upstairs bedroom. Immediately we can see eight aircraft in the area. With the BeagleBone Black now connected with Wi-Fi, we could even tuck the whole rig away and connect remotely to check in with a remote display.

BeagleBone Black Air Traffic Control Station

A number of programs for multiple operating systems take in ADS-B data and display maps that track the aircraft. You can put dump1090 in a mode where the data it is receiving is made available on the network and point one of these other programs to look for that data. A great, free program is adsbSCOPE. If you invoke dump1090 with the `--net` option, it will make the data available on port 3002. You can also invoke the `-aggressive` parameter, which will allow for the capture of more signals. Aggressive mode uses more CPU power, but we can afford that power.

```
root@beaglebone:~# dump1090 --interactive --net --aggressive
```

With the dump1090 program running happily on your BeagleBone Black, you can start the ADS-B software on a Windows PC. Some tutorials are available from the adsbSCOPE website that will help you configure the setup. Most importantly, you want to center on the actual location of your receiver. Play around with the settings in the Config menu for different results. You next want to configure to receive the aircraft information over the network. If you select Other, Network, Network Setup, you can configure the software to look for the BeagleBone Black at the network port on which the data is made available, 30002. This is shown in Figure 14.12.

You can then select Other, Network, RAW-data client Active, and the data from your BeagleBone Black will start streaming over Wi-Fi to this terminal, and you can watch aircraft activity in the area. Figure 14.13 shows the traffic in my local region over the course of a couple hours. Note that as of this book's writing, the ability of ADS-B to transmit position information is not required in 100% of commercial aircraft, but the number of aircraft equipped is increasing and the number of aircraft actually seen on the display is increasing in kind. Other countries have gotten to the point of 100% compliance, so afternoons and evenings on the East Coast of the United States can get pretty interesting with overseas flights. If you're a reader in a Europe country or another country where compliance is already mandatory, the screen can become very full. With just a basic setup and good antenna visibility, one can easily see aircraft as far as 85 kilometers. You now have a working air traffic control ground station, complete with graphical display! is adsbSCOPE.

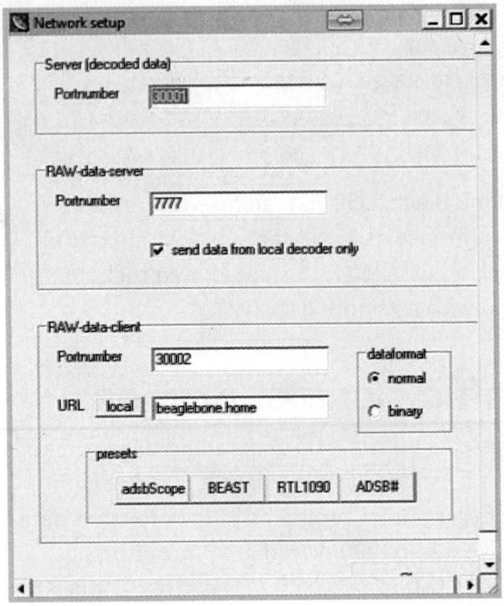

Network setup

Server (decoded data)
Portnumber 30001

RAW-data-server
Portnumber 7777
send data from local decoder only

RAW-data-client
Portnumber 30002
URL local beaglebone.home
dataformat
normal
binary
presets
adsbScope BEAST RTL1090 ADSB#

FIGURE 14.12 Configuring the adsbSCOPE network settings.

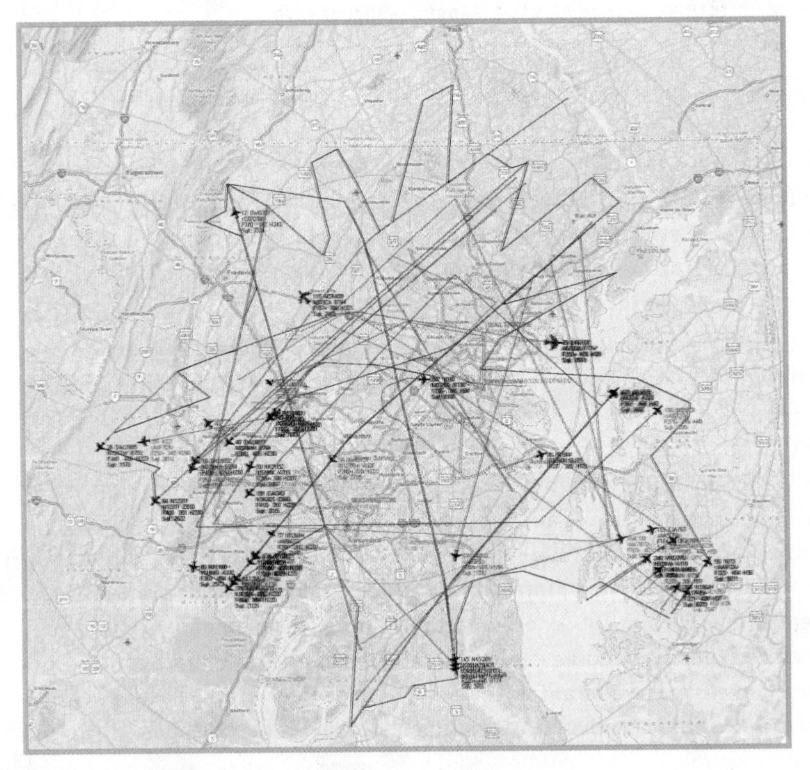

FIGURE 14.13 Local air traffic over several hours.

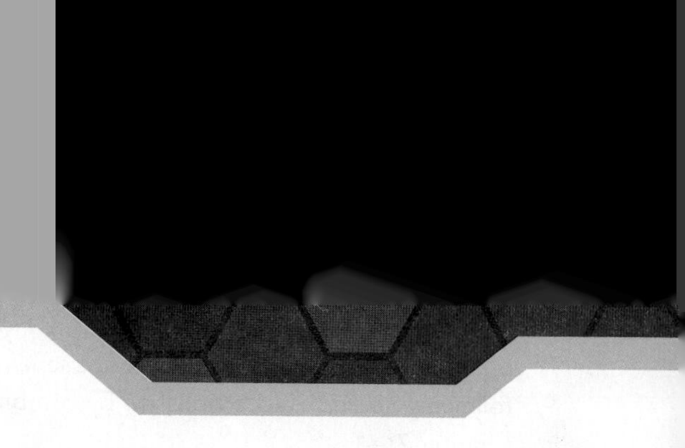

Moving Forward

Where do you go from here? The sky is the limit. The goal of this book has been to give you enough information to develop the basics of some really cool projects while also giving you the foundations to move forward to bigger and better projects, limited only by your imagination.

Chapter 1, "Embedded Computers and Electronics," assumed you had only a basic knowledge of computers and electronics. It introduced you to some of the basic concepts of a small computer and drew comparisons to computers with which you may be more familiar. You learned that a lot of power can be packed into a very small computer platform, and that you do not always need to rely on the massive power available in larger computers.

Chapter 2, "Introduction to the Hardware," introduced you to our board of choice: the BeagleBone Black. The history of the device and many technical aspects were covered in Chapter 2, giving you a better feel for what you would be working with as you continued in the book.

Chapter 3, "Getting Started," enabled you to start using the basics of the board and to execute a "Hello, World!" program. This included both the common computer programming version that, literally, prints out the text "Hello, World!" and the hardware version, which is a blinking LED. This project involved connecting to the BeagleBone Black via a built-in website and then using a terminal session. You also broke away from your USB connection to an Ethernet network.

In Chapter 4, "Hardware Basics," you delved deeper into the hardware specifications and what they mean. You also learned a lot about the very basics of electronics. This chapter may have required a couple passes if you were new to electronics or could have been skipped altogether if you have an electrical engineering degree.

Chapter 5, "A Little Deeper into Development," taught you the basics of computer programming. You implemented the blink code in a couple different languages and gained a better understanding of how these implementations differ beyond the blinking visible to the human eye. A great deal goes on under the covers in a program, and this chapter showed how you need to proceed if timing is important to your project. Chapter 5 was also the end of what could be considered the "basics" section of the book.

As you moved into Chapter 6, "Trying Other Operating Systems," you learned about the operating systems working with the hardware and handling a lot of interactions for you. You learned how to swap out the operating system in use and replace it with another. You used a specialized version of

Linux that runs a Super Nintendo emulator that can run games for which the user has legal access. This one operating system installation was basic, and because it was our first real project, we really only went into the basics for one operating system.

In Chapter 7, "Expanding the Hardware Horizon," you built a project to display information on an LED screen and learned about serial data communications. You saw that the following could be accomplished by combining what you learned in Chapter 6 and Chapter 7:

- Setting the serial display screen to show the user which game is currently being played
- Providing an update of how hard the CPU is working
- Displaying random, inspirational slogans to motivate the player

Chapter 8, "Low-Level Hardware and Capes," discussed the expansion environments available on the BeagleBone Black—that is, the Cape system. I cautioned you to fully read and understand any documentation used with a particular Cape. The Cape system is a shifting environment, and there are a number of ways to integrate a Cape into your project. Ensure that you select one for compatibility.

The SparkFun ProtoCape introduced in Chapter 8 presented an opportunity for you to interface further and combine the work of Chapters 6, 7, and 8. The serial display can be given its own interface to a connector designed into the ProtoCape, for example. The LCD display Capes discussed can be integrated to create a portable gaming solution!

Chapter 9, "Interacting with Your World, Part 1: Sensors," and Chapter 10, "Remote Monitoring and Data Collections," introduced the idea of sensing the environment around you, and we built a monitoring station to measure the environment in a room. We then jump into making things happen in our environment in Chapter 11, "Interacting with Your World, Part 2: Feedback and Actuators," with actuators and mechanisms.

Chapter 12, "Computer Vision," Chapter 13, "Sniffing Out Car Trouble," and Chapter 14, "Ground Control to Major Beagle," can be taken as capstone starter projects.

Project Ideas

Now that you are reading the final chapter of this book, you might be wondering where you can go from here. The remainder of this chapter gives you some ideas.

Portable Gaming Solutions

As mentioned previously, you can create a portable gaming environment. You don't need to use the SNES operating system introduced in Chapter 7 because plenty of decent Linux games are available. However, the SNES is a ton of fun—and nostalgic for those of us of a certain age.

You can combine the BeagleBone Black, Element 14 BB View LCD Cape, and SparkFun ProtoCape. You can also interface with the LCD display, as mentioned previously, to avoid carrying around a separate HDMI display. You can use the ProtoCape and what you learned about sensors in Chapter 9 to integrate actual SNES controllers. All of the electronics can be integrated onto the ProtoCape and transmitted down the GPIO available on the Cape stack. If you are worried about having enough GPIO available, you can use the Arduino-like environment on the CryptoCape to collect the button presses and transmit them to the BeagleBone Black over a serial protocol. Software can be written to handle these interactions and integrated to execute on startup, similar to the environment-monitoring project in Chapter 10.

Weather Station

You can expand greatly on the environmental monitoring station in Chapter 10 and turn it into a full-fledged weather station. For example, you could combine the following:

- BeagleBone Black
- Temperature sensor
- Light sensor
- Additional weather sensors from SparkFun Electronics (SEN-08942)
- The Wi-Fi dongle integrated in Chapter 14

With this combination, you can create a very useful weather station. You can continue to stream the information to the SparkFun data site, as in Chapter 10, but you could also create a website to execute on the BeagleBone Black itself to display the information. This way, anyone connected to your home network could just check the weather.

In-Car Computer

If you have a car without all the modern conveniences—but one that does have the OBD-II port used in Chapter 13—you can expand that setup into a full-featured car computer. The Artificial Intelligence to run Kit might be a bit much for the BeagleBone Black, but you can still get a lot of features, such as the following:

- BeagleBone Black
- SparkFun OBD-II to UART adapter
- Logic-level shifter
- Element14 BB View LCD Cape
- GPS sensor
- RTL-SDR

This provides an opportunity for a great number of applications. By adding in the LCD Cape, you gain a touchscreen. You can then create a graphical user environment that you can interact with and pull up all kinds of car parameters, or you can have a virtual instrument cluster for information that may not be made available on the normal instrument cluster.

Adding in the GPS, you can compare the car's reported speed against the GPS measurement, measure the heading, or just track the car's position. If you are trying to use one of those devices that rebroadcasts an iPod or other device over an FM signal but have trouble finding an open channel, you can integrate the RTL-SDR and an antenna to look for and report open sections of available spectrum.

More Advanced Aircraft "RADAR"

The aircraft-tracking system of Chapter 14 can be expanded a great deal with a few parts:

- BeagleBone Black
- USB hub
- RTL-SDR
- Element14 BB View LCD Cape
- GPS sensor
- Coaxial collinear antenna

By adding in a display and a GPS, you can create a version of our aircraft-tracking ground system that will update itself with the current position of the station. It is portable with the LCD Cape used for display, and if you build a coaxial collinear antenna, a quick Google search with ADS-B included will help, you can get a significant range. The basic setup in Chapter 14 allowed for reception as far away as 85 km, but well-positioned antennas have seen distances well in excess of 100 km.

Satellite Ground Station

The RTL-SDR can cover a great deal of the radio spectrum, and a very large amount of information is being transmitted over those waves. As of today, three satellites operated by the National Oceanic and Atmospheric Administration (NOAA) transmit the images of Earth they are capturing as a constant signal feed beneath their track. The satellites and their transmission frequency, as of this book's writing, are as follows:

- NOAA-15 (137.5MHz)
- NOAA-18 (137.913MHz)
- NOAA-19 (137.1MHz)

The parts you use for your satellite ground station, as with many other projects, will depend on how complex you want to go. Limited reception can be achieved with a basic antenna, but you can use Google to find other antenna designs. These NOAA images, known as Automatic Picture Transfer (APT), use a circularly polarized signal, whereas the signals we have looked at thus far with the RTL-SDR have all been linearly polarized. What polarization of a signal means can fill an entire book—actually, it has filled numerous books and online references and if you want to better understand what it means it is a fun area of signals to begin digging into.

With the following basics, you can achieve images like the one shown in Figure 15.1:

- BeagleBone Black
- USB hub
- RTL-SDR
- Wi-Fi dongle
- Basic antenna

FIGURE 15.1 Captured NOAA APT satellite image.

This is just one kind of signal you can receive from space. Here are some other signal types you can search for coming from satellites:

- International Space Station
- FUNCUBE 1
- Various CubeSat projects
- GPS

A little closer to Earth, you can look for the following:

- Air traffic cockpit communications (Narrow-Band FM)
- Car key dongle signals
- Ship position information
- Smart power meters
- Microwave ovens (2.4GHz)

Tools

The preceding section provided a list of just a few of the project ideas you can run with from the basics you learned in this book. With this in mind, you can never underestimate having a good toolkit at your disposal as you work on your projects. Here is a list of some of the key tools I used while developing this book:

- **Tektronix MSO2024B oscilloscope**—A powerful oscilloscope that provides a great deal of detail on signals. It's invaluable when you need to look closely at precision signals. The MSO2024B would be considered overkill for someone starting out, but the Tektronix TBS1000 series is great for people newbies and educational environments.
- **Oscium iMSO-104 and iMSO-204**—Another great oscilloscope solution, the Oscium scopes connect to an iPhone or iPad and turns the iOS device into an oscilloscope. Great for times when you can't lug a larger oscilloscope around.
- **Oscium LogiScope**—This easy-to-use logic analyzer from Oscium also utilizes iOS devices for the interface. I keep my Oscium gear in my go-bag to use at all times.
- **SparkFun digital multimeter (basic)**—This digital multimeter was part of a kit I received as a Father's day gift several years ago, and it's still an important part of my gear.
- **Hakko FX888D soldering station**—A good soldering iron is important. Many pins were soldered in the creation of this book.

This list is just some of the tools I used, but they are the big item tools I used repeatedly. Here are some other hand tools you'll want to consider:

- Diagonal cutters for wire snipping
- Wire strippers for removing insulation from wires
- Needle nose pliers
- An assortment of screwdrivers

Resources

You might benefit from the following on-line resources that I have used repeatedly not only in the development of this book and the projects in it, but in all of my making endeavors.

These websites have proven to be excellent references for better understanding many topics related to the projects in this book:

- **Getting started with the BeagleBoard**—http://beagleboard.org/getting-started

- **SparkFun tutorials**—http://learn.sparkfun.com
- **Adafruit tutorials**—http://learn.adafrui.com
- **BeagleBone Black GPIO interactive map**—http://eskimon.fr/beaglebone-black-gpio-interactive-map
- **Element14 community**—http://www.element14.com
- **Python tutorials**—https://docs.python.org/2/tutorial/
- **Explainshell**—http://explainshell.com/
  (This site shows what all the shell command arguments will do.)
- **Evil Mad Scientist Laboratories (EMSL) Electronics Basics**—http://www.evilmad-scientist.com/2013/basics-roundup/

The following three websites, mentioned throughout the book, are great sources for the add-on parts and electronics used in this book:

- **SparkFun**—http://sparkfun.com
- **Adafruit**—http://adafruit.com
- **Element14/Newark**—http://www.newark.com/?COM=element14_store_Home

No matter where your journey takes you, make sure it is one of discovery and fun. Good luck!

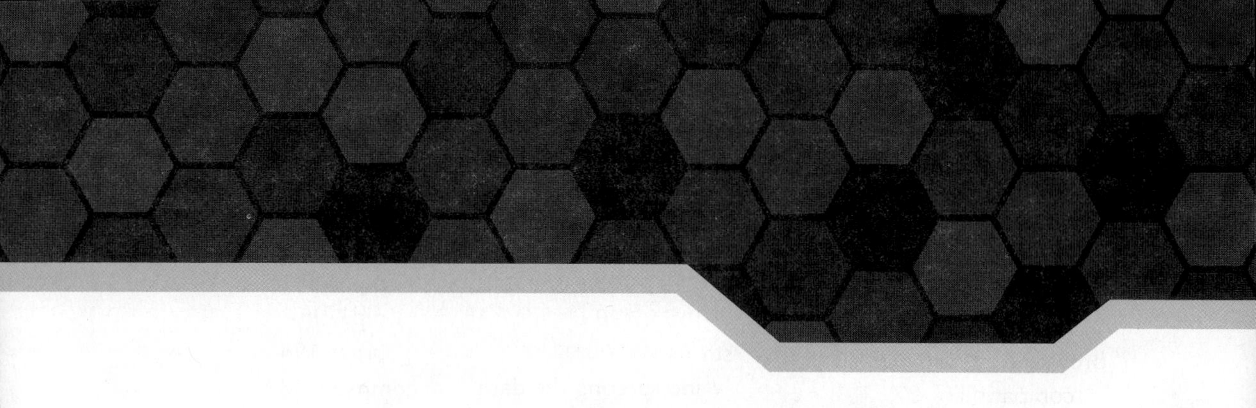

Index

Symbols

12-bit converter, 120

A

actuators, 149
Adafruit, 25
Adafruit_BBIO PWM library, 162
Adafruit Industries, 24
ADS-B
  adsbSCOPE, 223
  listening to aviation data, 221-223
adsbSCOPE, 223
aircraft tracking system, 228
alert.py project, 161
AM (amplitude modulation), 207
amps, 40-41
analog signals
  pulse sensors, 122-124
analog-to-digital converters, 120-121
  counts, 121
and operator, 87
aplay command, 218
APT (Automatic Picture Transfer), 228
apt-get command, 59-60
Arduino, 9
  Microcontroller, 10-11
arguments, 86
ARM
  TI Sitara processor, 20
ARM architecture, 15
array indexes, 89
Artificial Intelligence, 175
Atmel ATMega 328P, 11
aviation data, listening to, 221-223
  adsbSCOPE, 223

B

Babbage, Charles, 71
bandwidth, 217
Bash shell, 35
baud, 92
bbbservo.py project, 183-185
BB-View LCD Capes, 111
BeagleBoard, 15-16
BeagleBoard.org website, 73
BeagleBoard-xM, 16
BeagleBone, 17
  OpenROV project, 17-18
BeagleBone Black
  connecting to computer, 26
  Sitara processor, 19
  TI Sitara processor, 20
binary
  baud, 92
  bits, 81
    parity bit, 92
  bytes, 82-83
  counting in, 82
  hardware representation, 83-89
  LSB, 85
  MSB, 85
bits, 81-82
  baud, 92
  parity bit, 92
bitwise shift left, 86
BJT (bi-junction transistor), 151
blink.js, 30-32
blink.py project, 154
blinking lights
  implementing in Python, 62-65
bonescript, 30
  pinMode function, 31

BoneScript, 57
blink.js, 32
breadboard, 50
brightness of LEDs
controlling, 158
browsers
compatibility, 27
default website banner information, 27
building
LED circuit, 50-55
button circuits, 114
pull-down resistor, 114-115
pull-up resistor, 116
buttons
finger graphic, 150
bytes, 82-83

C

calculating
pulse durations, 162
calibrating the radio, 219-221
capacitors, 130
smoothing capacitors, 130
CapeManager, 108
Capes, 108, 111
BB-View LCD Capes, 111
CryptoCape, 110
ProtoCape, 109
capturing
photographs, 178-179
pictures, 173
video, 175
car computers, 189
MIL, 190
OBD, 190
OBD-II, 190
commands, 198
connecting to UART, 191-198
PIDs, 190
status
interpreting the data, 199-203
reading, 198-199
car_monitor.py program, 202-203
CascadeClassifier object, 181
cat command, 99
chmod command, 142-144
choosing
operating systems, 73
circuits
building, 50-55
clock, 7
RTC, 19
updating, 59
cmake command, 176, 215
code
commenting, 30
collectors, 151
command() method, 196
commands
aplay, 218
apt-get, 59-60
cat, 99
chmod, 142-144
cmake, 176, 216
echo, 98
git, 176
grep, 99
ipconfig, 35, 210
lsusb, 210
make, 216
man, 97-99
mkdir, 62
more, 99
OBD-II, 198
pip, 61
pipe, 99
print, 194
comments, 30
docstrings, 132-133
communications protocols
ADS-B
listening to aviation data, 221-223
communications_test.py project, 194
comparing
laptops and BeagleBone Black, 8
microcontrollers and microprocessors, 10
NPN and PNP transistors, 152
compatibility
web browsers, 27
compiled code, 65-69
computers
clock, 7
embedded computers, 5, 8
Arduino, 9
microcontroller, 10-11
GPIO ports, 8
Computer Vision, 175
connecting
board to computer, 26
to Ethernet, 32-37
to WiFi, 210-212
UART to OBD-II, 191, 197-198
webcam, 171

controlling
current, 152-154
LED brightness, 158
copying
image file, 74-75
counting in binary, 82
bits, 81
bytes, 82-83
hardware representation, 84-89
counts, 121
CPU clock, updating, 59
creating
portable gaming environment, 226
weather station, 227
CryptoCape, 110
current
controlling, 152-154
fading LEDs, 156-158
cycles per second, 124

D

datasheets, 118
dd, 75
decimal counting, 81
default website banner information, 27
demodulation, 212
desktops
comparing with BeagleBone Black, 8
Device Tree, 108
DHCP, 32
digital multimeters, 230
digital signals, 120
diodes, 48-50, 160
flow control, 160
LEDs, 48-49
circuit, building, 50-55
disk image, 73
copying, 74-75
distributions, 71
Ubuntu, 73
docstrings, 132-133
drivers
installing on your computer, 27
dump1090, 223
duration of pulses, calculating, 162
duty cycle, 157

E

EasyDriver, 165
echo command, 98
ECM
car status
interpreting the data, 199-203
reading, 198-199
ECM (Engine Control Module), 189
electricity
capacitors, 130
electromagnetic spectrum, 205
radio waves, 205-206
electronics
actuators, 149
amps, 40-41
diodes, 48-50
ground, 50
jumper wires, 52
oscilloscope, 54
resistance, 41-42
resistors, 45-48, 154
short circuit, 43-45
transducers, 113
transistors, 149-154
BJT, 151
LEDs, fading, 156-158
NPN, 151-152
PNP, 152
turn-on voltage, 152
voltage, 39-41
Watts, 40-41
work, 41-42
Element14 website, 66
embedded computers, 5, 8
Arduino, 9
microcontroller, 10-11
GPIO ports, 8
emitters, 151
eMMC, 74
ENIAC, 71
environment_monitor.py, 137
results, publishing, 137-138, 142
EOBD (European OBD), 190
Ethernet, 22
connecting, 32-33, 36-37

F

faces
identifying, 179-181
tracking, 182, 185, 188
face_tracker.py project, 180-181
fading LEDs, 156-158
file systems
GPIO
memory locations, 100-103
files, transferring, 62
FileZilla, 62
finger graphic, 150
flash memory, 22
flow control, 160

FM
bandwidth, 217
wideband FM, 217
FM (frequency modulation), 207
functions
(), 200
obd_read, 194
pinMode, 31
read_adc, 123
strip(), 201
wait for edge(), 116
functrions
arguments, 86
funtions, 86

G

gain, 217
gaming
portable gaming environment, creating, 226
git, 176
Git, 215
GitHub, 215
goal of this book, 225
GPIO, 23, 24
current, controlling, 152-154
headers, 23
memory locations, 100-103
pins, 102
default state for BeagleBone Black, 105, 108
mapping to GPIO memory locations, 100-102
mux, 103-105
slew rate, 105
GPIO (general-purpose input/output) ports, 5
GPIO ports, 8
grep command, 99
ground, 50

H

Hakko FX888D soldering station, 230
hardware
actuators, 149
Capes, 108, 111
BB-View LCD Capes, 111
CryptoCape, 110
ProtoCape, 109
digital multimeters, 230
diodes, 48-50
LED circuit, building, 50-55
LEDs
representing binary, 83-89
mux, 103-105
oscilloscope, 54
radio, testing, 216-219
resistors, 45-48
soldering irons, 230
voltage regulator, 94
webcam
connecting, 171
testing, 172
webcams
snapshot.py project, 173-175
GPIO, 23-24
headers, 23
processor, 20
RAM, 21, 22
hardware specifications for BeagleBone Black
Ethernet, 22
flash memory, 22
MicroSD, 22
headers, 23
heartbeats, 120
"Hello World!", 28-31
Hertz, 124
history
of Linux, 71-72

I

IDE (integrated development environment), 28
identifying
faces, 179-181
image file
copying, 74-75
images
capturing, 173, 178-179
faces, identifying, 180-181
faces, tracking, 182, 185, 188
in-car computer, 227-228
inspecting
UART, 93-96
insserv program, 144-146
installing
drivers, 27
Kalibrate, 219-220
OpenCV libraries, 175-176
operating system, 76-79
packages, 61
streamer, 172

Internet radio, 205
interpreted code, 57-58
  Python, 59-60
    blinking lights, implementing, 62-65
    functions, 86
interpreting
  OBD-II data, 199-203
int() funtion, 200
ipconfig command, 35, 210

J

JavaScript, 32
joysticks, 118
jumper wires, 52

K

Kalibrate
  installing, 219-220

L

laptops
  comparing with BeagleBone Black, 8
LBP (Local Binary Pattern), 180
LED circuit
  building, 50-55
LEDs, 48-49
  binary, representing, 83-89
  fading, 156-158
  polarity sensitive, 51
libraries
  Adafruit_BBIO PWM library, 162
  Git, 215
  libusb-1.0, 215
  OpenCV, 177-178
    installing, 175-176
  RCS, 215
  twisted, 61
libusb-1.0, 215
Linux
  Bash shell, 35
  cat command, 99
  distributions, 71
  echo command, 98
  grep command, 99
  history of, 71-72
  man command, 97-99
  more command, 99
  pipe command, 99
  redirects, 99
listening
  to aviation data, 221-223
    adsbSCOPE, 223
list indexes, 89
loading
  microSD card, 74-77
logic analyzers
  Oscium LogiScope, 230
logic-level converter, 192
logic states, 120
loss in waveform resolution, 125
LSB (least-significant bit), 85
lsusb command, 210

M

machine code, 57
make command, 216
man command, 97-99
mapping
  pins to GPIO memory locations, 100-102
memory
  flash memory, 22
  MicroSD external storage, 22
  RAM, 7
    in BeagleBone Black, 21-22
    paging, 22
  registers, 104
  volatile memory, 21
methods
  command(), 196
  speed(), 201
microcontrollers, 10-11
  Atmel ATMega 328P, 11
microprocessors
  ARM
    TI Sitara processor, 20
  ARM architecture, 15
  TI Sitara, 19
microSD card
  loading, 74-77
MicroSD external storage, 22
MIL (Malfunction Indicator Lamp), 190
mkdir command, 62
modes, 198
modifying
  permissions, 142-144
modulation, 207
  demodulation, 212
  phase modulation, 209
more command, 99
motors
  diodes, 160
  servo motors, 161-163
  stepper motors, 165-166
    stepper.py project, 167, 170
    winding inductance, 167

vibration motors, 159
alert.py project, 161
MSB (most-significant bit), 85
mux, 103-105

N

networking
Ethernet, 22
NOAA, 228
NPN transistors, 151-152
NTP (Network Time Protocol), 59
Nyquist sampling, 125

O

OBD-II, 190
car status
interpreting the data, 199-203
reading, 198-199
car status, reading, 198
commands, 198
connecting to UART, 191, 197-198
PIDs, 190
OBD (On-Board Diagnostics), 190
obd.py project, 195-196
obd_read function, 194
onboard computers, 189
ECM status
interpreting received data, 199-203
reading, 198-199
MIL, 190
OBD, 190
OBD-II, 190
commands, 198
connecting to UART, 191, 197-198
PIDs, 190
online resources, 230-231
on/off sensors, 113-116
OpenCV, 175
bbbservo.py project, 183-185
face_tracker.py project, 180-181
libraries, 177-178
libraries, installing, 175-176
photobooth.py project, 178-179
system installation, 177
tracker.py project, 185, 188
OpenCV (Open Computer Vision), 175
OpenROV project, 17-18
operating system
installing on BeagleBone Black, 76-79
operating systems
Linux
cat command, 99
distributions, 71
echo command, 98
grep command, 99
history of, 71-72
man command, 97-99
more command, 99
pipe command, 99
selecting, 73
VMS, 72
operators
and, 87
bitwise shift left, 86
oscilloscope, 54
oscilloscopes
Oscium iMSO-104, 230
Tektronix MSO2024B, 230
Oscium iMSO-104, 230
Oscium LogiScope, 230

P

packages, installing, 61
paging, 22
parallel communications
ribbon cables, 91
parity bit, 92
permissions
changing, 142-144
Phant, 138
phase modulation, 209
photobooth.py project, 178-179
photocells, 127, 134-136
photo_collection.py, 136
pictures
capturing, 173
PIDs, 198
PIDs (parameter IDs), 190
pinMode function, 31
pins, 102
default state, 105, 108
mapping to GPIO memory locations, 100-102
mux, 103-105
slew rate, 105
UART, 96
pip command, 61
pipe command, 99
PNP transistors, 152
polarity sensitive, 51

portable gaming environment,
 creating, 226
ported operating systems, 71
potentiometers, 118-120
power, 40-41
 short circuit, 43-45
ppm (parts per million), 219
print command, 194
print statement, 133
programming languages
 compiled code, 65-66, 69
 interpreted code, 57-58
 Python, 59-60
programs
 blink.js, 30
 car_monitor.py program, 202-203
 dd, 75
 dump1090, 223
 environment_monitor.py, 137
  results, publishing, 137-138, 142
 insserv, 144-146
 snapshot.py, 173-175
project ideas
 aircraft tracking system, 228
 in-car computer, 227-228
 portable gaming environment, 226
 satellite ground station, 228-230
 weather station, 227
projects
 bbbservo.py, 183-185
 blink.py, 154
 communications_test.py, 194
 face_tracker.py, 180-181
 obd.py, 195-196
 photo_collection.py, 136
 photobooth.py, 178-179
 pwm_blink.py, 156-157
 pwm_fade.py, 158
 servo.py, 162-163
 snapshot.py, 173-175
 tracker.py, 185, 188
 video.py, 177-178
ProtoCape, 109
publishing
 environment_monitor.py results, 137, 142
pull-down resistors, 114-115
pull-up resistors, 116
pulse sensors, 122-124
pushbutton circuit with LED indicator, 116-118
PuTTY, 33
PWM, 156-158
 Adafruit_BBIO PWM library, 162
 duty cycle, 157
 pulse durations, calculating, 162
pwm_blink.py, 156-157
pwm_fade.py, 158
PWM (pulse-width modulation), 156
Python, 59-60
 alert.py project, 161
 bbbservo.py project, 183-185
 binary counter program, 84-89
 blinking lights, implementing, 62-65
 blink.py project, 154
 chmod command, 142-144
 communications_test project, 194
 face_tracker.py project, 180-181
 functions, 86
 obd.py project, 195-196
 photobooth.py project, 178-179
 projects
  car_monitor.py, 202-203
  photo_collection.py, 136
  pwm_blink.py, 156-157
  pwm_fade.py, 158
  tmp36_collection.py, 130-132
 servo.py project, 162-163
 snapshot.py project, 173-175
 stepper.py project, 167, 170
 tracker.py project, 185, 188

R

radio
 AM, 207
 calibrating, 219-221
 FM, 207
  bandwidth, 217
 gain, 217
 Internet radio, 205
 listening to aviation data, 221-223
  adsbSCOPE, 223

phase modulation, 209
SDR, 212
RTL-SDR, 212-213
testing, 216-219
WiFi, 210-211
connecting to, 210-212
radio waves, 205-206
RAM, 7
in BeagleBone Black, 21-22
paging, 22
RCS (revision control system), 215
read_adc function, 123
reading
ECM status, 198-199
reconstructing waveforms, 125
redirects, 99
registers, 104
repository, 176
representing binary, 83-89
resistance, 41-42
short circuit, 43-45
resistors, 45-48, 154
ribbon cables, 91
rotation
sensing, 118-120
RTC (real-time clock), 19
RTL-SDR, 212-213
testing, 217-219

S

sample rates, 124-125
loss in waveform resolution, 125
Nyquist sampling, 125
sampling, 217
satellite ground station, 228-230
schematics
collectors, 151
emitters, 151
finger graphic, 150
SDR
listening to aviation data, 221-223
adsbSCOPE , 223
RTL-SDR, 212-213
SDR (software-defined radio), 212
security
permissions
changing, 142-144
selecting
operating systems, 73
sensors, 113-116
analog-to-digital converters, 120-121
buttons, 118
joysticks, 118
on/off sensors, 113
photocells, 127, 134-136
pulse sensors, 122-124
pushbutton circuit with LED indicator, 116-118
variable resistors, 118-120
serial communication
baud, 92
serial communications, 91
UART, 91-93
inspecting, 93-96
pins, 96
servo motors, 161-163
servo.py project, 162-163
shell scripts, 142
short circuit, 43-45
Sitara processor, 19-20
slew rate, 105
smoothing capacitors, 130
snapshot.py project, 173-175
SNES, 77
soldering irons
Hakko FX888D soldering station, 230
SparkFun
CryptoCape, 110
ProtoCape, 109
publishing environment_monitor.py results, 137-138, 142
SparkFun EasyDriver Stepper Motor Driver, 165
SparkFun Electronics, 25
SparkFun Pulse Sensor kit, 122-124
spectrum, 206
speed() method, 201
SSD (solid-state drive), 22
SSH
PuTTY, 33
SSH (Secure Shell), 33
start bits, 92
step angle, 165
stepper motors, 165-166
stepper.py project, 167, 170
winding inductance, 167
stepper.py project, 167, 170
stop bits, 92
streamer, 172
installing, 172
strip() function, 201
suppliers, 25
system disk
microSD card
loading, 74-77

T

Tektronix MSO2024B oscilloscope, 230
testing
  radio hardware, 216-219
  webcam, 172
TI Sitara processor, 19-20
tmp36_collection.py, 130-132
Torvalds, Linus, 72
tracker.py project, 185-188
tracking
  faces, 182-188
transducers, 113
  sensors
    analog-to-digital converters, 120-121
    buttons, 118
    joysticks, 118
    pulse sensors, 122-124
    variable resistors, 118-120
transferring files, 62
transistors, 49, 149-154
  BJT, 151
  LEDs, fading, 156-158
  NPN, 151-152
  PNP, 152
  turn-on voltage, 152
truth tables, 87
turn-on voltage, 152
twisted library, 61

U

UART
  baud, 92
  connecting to OBD-II, 191, 197-198
  parity bit, 92
  pins, 96
UART (Universal Asynchronous, 93-96
UART (Universal Asynchronous Receiver/Transmitter), 91-93
Ubuntu, 73
Unix, 72
updating the clock, 59
USB Audio Adapter (Adafruit), 218
user lights, 26
utilities
  cmake, 215
  rtl_fm, 217

V

variable resistors, 118-120
vibration motors, 159
  alert.py project, 161
video
  capturing, 175
video camera
  connecting, 171
  testing, 172
video cameras
  OpenCV
    bbbservo.py project, 183-185
    face_tracker.py project, 180-181
    libraries, 177-178
    library installation, 176
    photobooth.py project, 178-179
    system installation, 177
    tracker.py project, 185, 188
video.py project, 177-178
VMS, 72
volatile memory, 21
voltage, 39-41
  brightness, 136
  capacitors, 130
  turn-on voltage, 152
voltage divider circuit, 119
voltage regulator, 94

W

wait for edge() function, 116
Watts, 40-41
waveforms
  loss in resolution, 125
  reconstructing, 125
weather station
  creating, 227
web browsers
  compatibility, 27
  default website banner information, 27
webcam
  connecting, 171-172
webcams
  snapshot.py project, 173-175
websites
  Adafruit, 25
  BeagleBoard.org, 73
  Element14, 66
  GitHub, 215
  online resources, 230-231
  SparkFun Electronics, 25
wideband FM, 217
WiFi, 210-211
  connecting to, 210-212
Win32 Disk Imager, 74
winding inductance, 167
work, 41-42

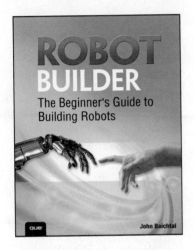

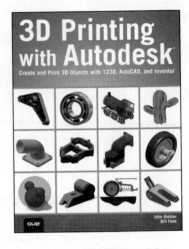